T0224039

SpringerBriefs in Mathematics

Series editors

Nicola Bellomo
Michele Benzi
Palle E.T. Jorgensen
Tatsien Li
Roderick Melnik
Lothar Reichel
Otmar Scherzer
Benjamin Steinberg
Yuri Tschinkel
G. George Yin
Ping Zhang

SpringerBriefs in Mathematics showcases expositions in all areas of mathematics and applied mathematics. Manuscripts presenting new results or a single new result in a classical field, new field, or an emerging topic, applications, or bridges between new results and already published works, are encouraged. The series is intended for mathematicians and applied mathematicians.

More information about this series at http://www.springer.com/series/10030

Tatsien Li · Ke Wang · Qilong Gu

Exact Boundary Controllability of Nodal Profile for Quasilinear Hyperbolic Systems

 Springer

Tatsien Li
School of Mathematical Sciences
Fudan University
Shanghai
China

Qilong Gu
School of Mathematical Sciences
Shanghai Jiao Tong University
Shanghai
China

Ke Wang
College of Science
Donghua University
Shanghai
China

ISSN 2191-8198   ISSN 2191-8201   (electronic)
SpringerBriefs in Mathematics
ISBN 978-981-10-2841-0   ISBN 978-981-10-2842-7   (eBook)
DOI 10.1007/978-981-10-2842-7

Library of Congress Control Number: 2016954709

Mathematics Subject Classification (2010): 93B05, 35L50, 35L05, 35L72

Printed on acid-free paper

This Springer imprint is published by Springer Nature
The registered company is Springer Nature Singapore Pte Ltd.
The registered company address is: 152 Beach Road, #22-06/08 Gateway East, Singapore 189721, Singapore

Preface

The exact boundary controllability of hyperbolic systems is of great importance in both theory and applications. A complete theory on the local exact boundary controllability for 1-D quasilinear hyperbolic systems has been established by means of a constructive method with modular structure, based on three basic ingredients: existence and uniqueness of semi-global classical solution to the mixed initial-boundary value problem; exchanging the role of the time variable t and the space variable x; and uniqueness of classical solution to the one-sided mixed initial-boundary value problem (see Li [12]).

Recently, stimulated by some practical applications, M. Gugat et al. proposed a new kind of exact boundary controllability in [10]. Their initiative was almost immediately generalized to general 1-D first order quasilinear hyperbolic systems with general nonlinear boundary conditions [13]. Differently from the usual exact boundary controllability that asks the solution to the system under certain boundary controls to meet a given final state at a suitably large time $t = T$, this kind of exact boundary controllability, called the exact boundary controllability of nodal profile, requires that the value of solution satisfies the given profiles on one or several nodes for $t \geq T$ by means of boundary controls.

Although the exact boundary controllability of nodal profile is not a special example of the usual exact boundary controllability in 1-D case, in this book we will show that the previous three basic ingredients can be still elegantly used to give a constructive method with modular structure to deal with the local exact boundary controllability of nodal profile in a systematic way, not only on a single interval, but also on a tree-like network with general topology.

All preliminaries are given in the first three chapters for completeness, and then, the next four chapters will present the corresponding exact boundary controllability of nodal profile for 1-D first order quasilinear hyperbolic systems or 1-D quasilinear wave equations on a single interval or a tree-like network with general topology, respectively.

The basic contents given in this book come essentially from some related papers of the authors (see Li [13], Gu and Li [9], Wang [21], and Wang and Gu [22]) with necessary supplements and modifications (see especially Chap. 3, Sects. 4.5–4.6,

Sect. 5.3, Sects. 5.7–5.8, Sect. 6.5, and Sects. 7.5–7.6). The draft of Chaps. 2, 6, and 7 was compiled by Ke Wang and that of Chaps. 1, 3–5 by Tatsien Li. The final version of the whole book was revised and finalized by Tatsien Li.

Shanghai, China   Tatsien Li
                                                                           Ke Wang
                                                                           Qilong Gu

Contents

Chapter 1
First Order Quasilinear Hyperbolic Systems

1.1 1-D First Order Quasilinear Hyperbolic Systems

We consider the following 1-D first order quasilinear system

$$\frac{\partial u}{\partial t} + A(u)\frac{\partial u}{\partial x} = F(u), \tag{1.1}$$

where t is the time variable, x is the spatial variable, $u = (u_1, \ldots, u_n)^T$ is the unknown vector function of (t, x), $A(u)$ is a given $n \times n$ matrix with suitably smooth entries $a_{ij}(u)$ $(i, j = 1, \ldots, n)$, and $F(u) = (f_1(u), \ldots, f_n(u))^T$ is a given vector function of u with suitably smooth elements.

Definition 1.1 (1.1) is a **hyperbolic system**, if, for any given u on the domain under consideration, the matrix $A(u)$ has n real eigenvalues $\lambda_1(u), \ldots, \lambda_n(u)$ and a complete set of left (resp. right) eigenvectors $l_i(u) = (l_{i1}(u), \ldots, l_{in}(u))$ (resp. $r_i(u) = (r_{1i}(u), \ldots, r_{ni}(u))^T$) $(i = 1, \ldots, n)$:

$$l_i(u)A(u) = \lambda_i(u)l_i(u) \tag{1.2}$$

and

$$A(u)r_i(u) = \lambda_i(u)r_i(u). \tag{1.3}$$

We have

$$\det |l_{ij}(u)| \neq 0 \quad (\text{resp. } \det |r_{ij}(u)| \neq 0). \tag{1.4}$$

Without loss of generality, we may assume that on the domain under consideration

$$l_i(u)r_j(u) \equiv \delta_{ij} \quad (i, j = 1, \ldots, n) \tag{1.5}$$

© The Author(s) 2016
T. Li et al., *Exact Boundary Controllability of Nodal Profile
for Quasilinear Hyperbolic Systems*, SpringerBriefs in Mathematics,
DOI 10.1007/978-981-10-2842-7_1

and

$$r_i^T(u)r_i(u) \equiv 1 \quad (i = 1, \dots, n), \tag{1.6}$$

where δ_{ij} stands for the Kronecker symbol.

Let

$$L(u) = \begin{pmatrix} l_1(u) \\ \vdots \\ l_n(u) \end{pmatrix} = (l_{ij}(u)) \tag{1.7}$$

and

$$R(u) = (r_1(u), \dots, r_n(u)) = (r_{ij}(u)) \tag{1.8}$$

be the matrices composed of the left and right eigenvectors, respectively, and let

$$\Lambda(u) = \mathrm{diag}\{\lambda_1(u), \dots, \lambda_n(u)\}. \tag{1.9}$$

By (1.2) and (1.3), we have

$$L(u)A(u) = \Lambda(u)L(u) \tag{1.10}$$

and

$$A(u)R(u) = R(u)\Lambda(u). \tag{1.11}$$

Moreover, by (1.5) we have

$$L(u)R(u) = I, \tag{1.12}$$

where I is the $n \times n$ unit matrix, namely, we have

$$R(u) = L^{-1}(u). \tag{1.13}$$

Definition 1.2 If, for any given u on the domain under consideration, the matrix $A(u)$ has n distinct real eigenvalues

$$\lambda_1(u) < \lambda_2(u) < \cdots < \lambda_n(u), \tag{1.14}$$

then, the set of left (resp. right) eigenvectors must be complete. This kind of hyperbolic system (1.1) is called a **strictly hyperbolic system**.

It is easy to see that, for any strictly hyperbolic system, all $\lambda_i(u)$, $l_i(u)$ and $r_i(u)$ $(i = 1, \ldots, n)$ have the same regularity as $A(u)$. In what follows, for general hyperbolic systems, we always assume that all $\lambda_i(u)$, $l_i(u)$ and $r_i(u)$ $(i = 1, \ldots, n)$ have the same regularity as $A(u)$.

Remark 1.1 There are many physical models such as traffic flow, 1-D gas dynamics without viscosity, 1-D elastic waves, blood flow, supply chains, irrigation channels, chromatography etc. which are all described by 1-D quasilinear hyperbolic systems. In particular, many physical models mentioned above are considered on networks (cf. [1, 3, 5, 12] and the references cited there). However, in this book the main topic will be treated only for certain well-chosen models, similar discussions can be done in principle for other models.

1.2 Characteristic Form of Hyperbolic System

Multiplying the i-th left eigenvector $l_i(u)$ on system (1.1) from the left and noting (1.2), we get

$$l_i(u)\left(\frac{\partial u}{\partial t} + \lambda_i(u)\frac{\partial u}{\partial x}\right) = \tilde{F}_i(u) \stackrel{\text{def.}}{=} l_i(u)F(u) \quad (i = 1, \ldots, n). \tag{1.15}$$

Noting (1.4), system (1.15) is equivalent to system (1.1).

The i-th equation of system (1.15) consists of only the directional derivatives of the unknown vector function $u = u(t, x)$ along the i-th characteristic

$$\frac{dx}{dt} = \lambda_i(u). \tag{1.16}$$

System (1.15) is then called the **characteristic form** of hyperbolic system (1.1).

In the special case that $l_i(u)$ $(i = 1, \ldots, n)$ are all independent of u: $l_i(u) = l_i$ $(i = 1, \ldots, n)$, setting

$$v_i = l_i u \quad (i = 1, \ldots, n), \tag{1.17}$$

system (1.15) can be reduced to a system of diagonal form with respect to the variables v_i $(i = 1, \ldots, n)$ as follows:

$$\frac{\partial v_i}{\partial t} + \bar{\lambda}_i(v)\frac{\partial v_i}{\partial x} = \bar{\tilde{F}}_i(v) \quad (i = 1, \ldots, n), \tag{1.18}$$

where $\bar{\lambda}_i(v) = \lambda_i(u)$ and $\bar{\tilde{F}}_i(v) = \tilde{F}_i(u)$, since (1.17) is an invertible transformation from u to v. System (1.18) can be written in the following matrix form:

$$\frac{\partial v}{\partial t} + \bar{A}(v)\frac{\partial v}{\partial x} = \bar{\bar{F}}(v), \tag{1.19}$$

in which $v = (v_1, \ldots, v_n)^T$,

$$\bar{A}(v) = \text{diag}\{\tilde{\lambda}_1(v), \ldots, \tilde{\lambda}_n(v)\} \tag{1.20}$$

is a diagonal matrix, and $\bar{\bar{F}}(v) = (\bar{\bar{F}}_1(v), \ldots, \bar{\bar{F}}_n(v))^T$.

From this point of view, in what follows

$$v_i = l_i(u)u \quad (i = 1, \ldots, n) \tag{1.21}$$

will be called the **diagonal variables**.

1.3   Reducible Quasilinear Hyperbolic System. Riemann Invariants

We now consider a special situation, the so-called **reducible quasilinear hyperbolic system** which consists of two equations with $F(u) \equiv 0$.

The corresponding characteristic form is then

$$l_i(u)\left(\frac{\partial u}{\partial t} + \lambda_i(u)\frac{\partial u}{\partial x}\right) = 0, \quad (i = 1, 2). \tag{1.22}$$

It is well-known that at least in a local domain of $u = (u_1, u_2)^T$, there exist integral factors $\Pi_i(u) \neq 0$ $(i = 1, 2)$, such that

$$\Pi_i(u)l_i(u)du = \Pi_i(u)(l_{i1}(u)du_1 + l_{i2}(u)du_2)$$

is a total differential dr and ds with $r = r(u_1, u_2)$ and $s = s(u_1, u_2)$, respectively. Hence, taking r and s as new unknown functions, (1.22) equivalently reduces to

$$\begin{cases} \dfrac{\partial r}{\partial t} + \lambda_1(r, s)\dfrac{\partial r}{\partial x} = 0, \\[2mm] \dfrac{\partial s}{\partial t} + \lambda_2(r, s)\dfrac{\partial s}{\partial x} = 0. \end{cases} \tag{1.23}$$

This is a system of diagonal form, from which it is easy to see that r is a constant along each first characteristic determined by

$$\frac{dx}{dt} = \lambda_1(r, s), \tag{1.24}$$

while, s is a constant along each second characteristic determined by

$$\frac{dx}{dt} = \lambda_2(r, s). \tag{1.25}$$

r and s are then called the **Riemann invariants**.

1.4 Saint-Venant System for Unsteady Flows on a Single Open Canal

We now consider the Saint-Venant system for unsteady flows on a single open canal (cf. [4]).

Suppose that the canal is horizontal and cylindrical, the corresponding system can be written as

$$\begin{cases} \dfrac{\partial A}{\partial t} + \dfrac{\partial (AV)}{\partial x} = 0, \\[2mm] \dfrac{\partial V}{\partial t} + \dfrac{\partial S}{\partial x} = 0, \end{cases} \tag{1.26}$$

where $A = A(t, x)$ stands for the area of the cross section of the canal at x occupied by the water at time t, $V = V(t, x)$ the average velocity over the cross section and

$$S = \frac{1}{2}V^2 + gh(A) + gY_b, \tag{1.27}$$

where g is the gravity constant, constant Y_b denotes the altitude of the bed of canal and

$$h = h(A) \tag{1.28}$$

is the depth of the water, $h(A)$ being a suitably smooth function of A, such that

$$h'(A) > 0. \tag{1.29}$$

We say that any given equilibrium state $(A, V) = (A_0, V_0)$ of system (1.26) with $A_0 > 0$ belongs to the **subcritical case**, if

$$|V_0| < \sqrt{gA_0h'(A_0)}. \tag{1.30}$$

Thus, it is easy to see that, in a neighbourhood of any given subcritical equilibrium state $(A, V) = (A_0, V_0)$, (1.26) is a strictly hyperbolic system with two distinct real eigenvalues

$$\lambda_1 \stackrel{\text{def.}}{=} V - \sqrt{gAh'(A)} < 0 < \lambda_2 \stackrel{\text{def.}}{=} V + \sqrt{gAh'(A)} \tag{1.31}$$

and the corresponding left eigenvectors can be taken as

$$l_1 = \left(-\sqrt{\frac{gh'(A)}{A}}, 1\right), \quad l_2 = \left(\sqrt{\frac{gh'(A)}{A}}, 1\right). \tag{1.32}$$

Noting (1.31) and (1.32), the corresponding characteristic form of Saint-Venant system (1.26) can be written as

$$\left\{ \begin{array}{l} -\sqrt{\dfrac{gh'(A)}{A}} \left(\dfrac{\partial A}{\partial t} + \lambda_1 \dfrac{\partial A}{\partial x}\right) + \left(\dfrac{\partial V}{\partial t} + \lambda_1 \dfrac{\partial V}{\partial x}\right) = 0, \\[2mm] \sqrt{\dfrac{gh'(A)}{A}} \left(\dfrac{\partial A}{\partial t} + \lambda_2 \dfrac{\partial A}{\partial x}\right) + \left(\dfrac{\partial V}{\partial t} + \lambda_2 \dfrac{\partial V}{\partial x}\right) = 0. \end{array} \right. \tag{1.33}$$

Introducing the Riemann invariants r and s as follows:

$$\left\{ \begin{array}{l} 2r = V - V_0 - G(A), \\ 2s = V - V_0 + G(A), \end{array} \right. \tag{1.34}$$

where

$$G(A) = \int_{A_0}^{A} \sqrt{\frac{gh'(A)}{A}} dA, \tag{1.35}$$

we have

$$\left\{ \begin{array}{l} V = r + s + V_0, \\ A = H(s - r) > 0, \end{array} \right. \tag{1.36}$$

where H is the inverse function of $G(A)$,

$$H(0) = A_0 \tag{1.37}$$

and

$$H'(0) = \sqrt{\frac{A_0}{gh'(A_0)}} > 0. \tag{1.38}$$

Taking (r, s) as new unknown variables, Saint-Venant system (1.26) reduces to the system (1.23) of diagonal form, in which

$$\left\{ \begin{array}{l} \lambda_1(r, s) = r + s + V_0 - \sqrt{gH(s - r)h'(H(s - r))} < 0, \\ \lambda_2(r, s) = r + s + V_0 + \sqrt{gH(s - r)h'(H(s - r))} > 0. \end{array} \right. \tag{1.39}$$

1.5 Semi-global C^1 Solutions to the Mixed Initial-Boundary Value Problem

A systematic theory on the local C^1 solution to the mixed initial-boundary value problem for the first order quasilinear hyperbolic system (1.1) can be found in Li and Yu [20]. In order to study the local exact boundary controllability of nodal profile for first order quasilinear hyperbolic systems, it is necessary to consider the semi-global C^1 solutions in a neighbourhood of an equilibrium, i.e., the C^1 solution on the time interval $0 \le t \le T_0$, where $T_0 > 0$ is a preassigned and possibly quite large number.

For the first order quasilinear hyperbolic system (1.1), we assume that

$$F(0) = 0. \tag{1.40}$$

Then, $u = 0$ is an equilibrium of the system (1.1).

We first consider the case that on the domain under consideration, there are no zero eigenvalues:

$$\lambda_r(u) < 0 < \lambda_s(u) \quad (r = 1, \ldots, m; s = m+1, \ldots, n). \tag{1.41}$$

We give the initial condition

$$t = 0: \quad u = \varphi(x), \quad 0 \le x \le L, \tag{1.42}$$

where L is the length of the spatial interval, and the following boundary conditions:

$$x = 0: \quad v_s = G_s(t, v_1, \ldots, v_m) + H_s(t) \quad (s = m+1, \ldots, n), \tag{1.43}$$

$$x = L: \quad v_r = G_r(t, v_{m+1}, \ldots, v_n) + H_r(t) \quad (r = 1, \ldots, m), \tag{1.44}$$

where v_i $(i = 1, \ldots, n)$ are the diagonal variables given by (1.21), and without loss of generality, we assume that

$$G_i(t, 0, \ldots, 0) \equiv 0 \quad (i = 1, \ldots, n). \tag{1.45}$$

We point out that (1.43) and (1.44) are the most general boundary conditions to guarantee the well-posedness of the forward mixed initial-boundary value problem (1.1), (1.42), (1.43) and (1.44).

The characteristics of system (1.1) are determined by

$$\frac{dx}{dt} = \lambda_i(u) \quad (i = 1, \ldots, n),$$

in which $u = u(t, x)$ is a C^1 solution to system (1.1). The characteristics coming to the boundary $x = 0$ (resp. $x = L$) from the domain $\{(t, x)|t \ge 0, 0 \le x \le L\}$ are called the **incoming characteristics** on $x = 0$ (resp. $x = L$), while, the characteristics departing from the boundary $x = 0$ (resp. $x = L$) to the domain $\{(t, x)|t \geq 0, 0 \leq x \leq L\}$ are called the **outgoing characteristics** on $x = 0$ (resp. $x = L$).

Thus, the characters of the boundary conditions (1.43) and (1.44) can be expressed as

(1) The number of the boundary conditions on $x = 0$ (resp. $x = L$) is equal to the number of incoming characteristics on it.
(2) The boundary conditions on $x = 0$ (resp. $x = L$) are written in the form that the diagonal variables $v_s (s = m + 1, \ldots, n)$ (resp. $v_r (r = 1, \ldots, m)$) corresponding to the incoming characteristics are explicitly expressed by other diagonal variables.

By means of an extension method of local C^1 solution to the mixed initial-boundary value problem, we have (cf. [12, 15])

Lemma 1.1 *Suppose that $l_{ij}(u)$, $\lambda_i(u)$, $f_i(u)$, $G_i(t, \cdot)$, $H_i(t)$ $(i, j = 1, \ldots, n)$ and $\varphi(x)$ are all C^1 functions with respect to their arguments. Suppose furthermore that (1.40), (1.41) and (1.45) hold. Suppose finally that the conditions of C^1 compatibility (see Remark 1.2 below) are satisfied at the points $(t, x) = (0, 0)$ and $(0, L)$, respectively. Then for any given $T_0 > 0$, the forward mixed initial-boundary value problem (1.1), (1.42), (1.43) and (1.44) admits a unique semi-global C^1 solution $u = u(t, x)$ with small C^1 norm on the domain*

$$R(T_0) = \{(t, x)|0 \leq t \leq T_0, 0 \leq x \leq L\}, \tag{1.46}$$

provided that $\|\varphi\|_{C^1[0,L]}$ and $\|H\|_{C^1[0,T_0]}$ are suitably small (depending on T_0).

Remark 1.2 The **conditions of C^1 compatibility** at the point $(t, x) = (0, 0)$ for the forward mixed initial-boundary value problem (1.1), (1.42), (1.43) and (1.44) can be obtained as follows:

At the point $(t, x) = (0, 0)$, from the initial condition (1.42) we have

$$u(0, 0) = \varphi(0), \tag{1.47}$$

which should satisfy the boundary condition (1.43) at that point, then we get the following **conditions of C^0 compatibility** at the point $(t, x) = (0, 0)$:

$$v_s(\varphi(0)) = G_s(0, v_1(\varphi(0)), \ldots, v_m(\varphi(0))) + H_s(0) \quad (s = m + 1, \ldots, n), \tag{1.48}$$

in which $v_i = l_i(u)u$ $(i = 1, \ldots, n)$.

We now drive the conditions of C^1 compatibility at the point $(t, x) = (0, 0)$. From the initial condition (1.42), we have

$$u_x(0, 0) = \varphi'(0). \tag{1.49}$$

Then, by means of system (1.1), we get from (1.47) and (1.49) that

$$u_t(0,0) = F(\varphi(0)) - A(\varphi(0))\varphi'(0). \tag{1.50}$$

Taking the differentiation of the boundary conditions (1.43) with respect to t, the resulting formulas should be satisfied at the point $(t, x) = (0, 0)$ by $u(0, 0)$ and $u_t(0, 0)$ given by (1.47) and (1.50). These are the conditions of C^0 compatibility of the first order partial derivatives of the solution at the point $(t, x) = (0, 0)$, which together with the conditions of C^0 compatibility (1.48) give the conditions of C^1 compatibility at the point $(t, x) = (0, 0)$ for the forward mixed problem under consideration. We omit the details here.

The conditions of C^1 compatibility at the point $(t, x) = (0, L)$ can be obtained in a similar way.

We now consider the case that on the domain under consideration, there are zero eigenvalues:

$$\lambda_p(u) < \lambda_q(u) \equiv 0 < \lambda_r(u)$$
$$(p = 1, \ldots, l; q = l+1, \ldots, m; r = m+1, \ldots, n). \tag{1.51}$$

Under this assumption, since the incoming characteristics are λ_r $(r = m+1, \ldots, n)$ (resp. λ_p $(p = 1, \ldots, l)$) on $x = 0$ (resp. $x = L$), while the outgoing characteristics are λ_p $(p = 1, \ldots, l)$ (resp. λ_r $(r = m+1, \ldots, n)$) on $x = 0$ (resp. $x = L$), the previous boundary conditions (1.43) and (1.44) should be replaced by

$$x = 0: \quad v_r = G_r(t, v_1, \ldots, v_l, v_{l+1}, \ldots, v_m) + H_r(t)$$
$$(r = m+1, \ldots, n), \tag{1.52}$$

$$x = L: \quad v_p = G_p(t, v_{l+1}, \ldots, v_m, v_{m+1}, \ldots, v_n) + H_p(t)$$
$$(p = 1, \ldots, l) \tag{1.53}$$

with

$$G_p(t, 0, \ldots, 0) \equiv G_r(t, 0, \ldots, 0) \equiv 0$$
$$(p = 1, \ldots, l; r = m+1, \ldots, n). \tag{1.54}$$

Similarly to Lemma 1.1, we have (cf. [12])

Lemma 1.2 *Suppose that $l_i, \lambda_i, F, G_p, G_r, H_p, H_r$ ($i = 1, \ldots, n$; $p = 1, \ldots, l$; $r = m+1, \ldots, n$) and φ are all C^1 functions with respect to their arguments. Suppose furthermore that (1.40), (1.51) and (1.54) hold. For any given and possibly quite large $T_0 > 0$, if $\|\varphi\|_{C^1[0,L]}$ and $\|(H_1, \ldots, H_l, H_{m+1}, \ldots, H_n)\|_{C^1[0,T_0]}$ are sufficient small (depending on T_0), and the conditions of C^1 compatibility are satisfied at the points $(t, x) = (0, 0)$ and $(0, L)$, respectively, then the forward* *mixed initial-boundary value problem* (1.1), (1.42), (1.52) *and* (1.53) *admits a unique semi-global* C^1 *solution* $u = u(t, x)$ *with small* C^1 *norm on the domain* $R(T_0) = \{(t, x)|0 \le t \le T_0, 0 \le x \le L\}$.

Remark 1.3 For the backward mixed initial-boundary value problem of the system (1.1) with the final condition

$$t = T_0 : \quad u = \Phi(x), \tag{1.55}$$

since on the boundary $x = 0$, the incoming characteristics λ_s $(s = m + 1, \ldots, n)$ in the case (1.41) or λ_r $(r = m + 1, \ldots, n)$ in the case (1.51) all become the outgoing characteristics, while the outgoing characteristics λ_r $(r = 1, \ldots, m)$ in the case (1.41) or λ_p $(p = 1, \ldots, l)$ in the case (1.51) all become the incoming characteristics; and on the boundary $x = L$, similar situations happen, we should prescribe the boundary conditions (in the case (1.41))

$$x = 0 : \quad v_r = \widetilde{G}_r(t, v_{m+1}, \ldots, v_n) + \widetilde{H}_r(t) \quad (r = 1, \ldots, m), \tag{1.56}$$

$$x = L : \quad v_s = \widetilde{G}_s(t, v_1, \ldots, v_m) + \widetilde{H}_s(t) \quad (s = m + 1, \ldots, n), \tag{1.57}$$

or the boundary conditions (in the case (1.51))

$$x = 0 : \quad v_p = \widetilde{G}_p(t, v_{l+1}, \ldots, v_m, v_{m+1}, \ldots, v_n) + \widetilde{H}_p(t) \quad (p = 1, \ldots, l), \tag{1.58}$$

$$x = L : \quad v_r = \widetilde{G}_r(t, v_1, \ldots, v_l, v_{l+1}, \ldots, v_m) + \widetilde{H}_r(t) \quad (r = m + 1, \ldots, n), \tag{1.59}$$

Similar results as in Lemmas 1.1 and 1.2 hold.

1.6   Exchanging the Role of t and x

In the case that there are no zero eigenvalues (see (1.41)), we can exchange the role of t and x, and the original system (1.1) is rewritten as

$$\frac{\partial u}{\partial x} + A^{-1}(u)\frac{\partial u}{\partial t} = \widetilde{F}(u) \stackrel{\text{def.}}{=} A^{-1}(u)f(u) \tag{1.60}$$

with

$$\widetilde{F}(0) = 0. \tag{1.61}$$

Thus, the matrix $A(u)$ is replaced by its inverse $A^{-1}(u)$, then it is easy to get

Lemma 1.3 *Under the assumption* (1.41), *one can exchange the role of* t *and* x *so that the eigenvalues* $\lambda_i(u)$ $(i = 1, \ldots, n)$ *become* $1/\lambda_i(u)$ $(i = 1, \ldots, n)$, *while,* the left eigenvectors $l_i(u)$ $(i = 1, \ldots, n)$ keep unchanged, then the variables v_i $(i = 1, \ldots, n)$ are still given by the same formula (1.21).

Thus, we can consider the corresponding rightward (resp. leftward) mixed initial-boundary value problem and we have similar results as in Lemma 1.1.

1.7 Uniqueness of C^1 Solution to the One-Sided Mixed Initial-Boundary Value Problem

First we consider the one-sided forward mixed initial-boundary value problem for the system (1.1), the initial condition (1.42) and the boundary condition (1.52) on $x = 0$ under the assumption (1.51) ((1.41) can be regarded as the special case of (1.51) in which we take $l = m$).

Based on the finite speed of propagation, we have (cf. [20])

Lemma 1.4 *Under the assumption* (1.51), *the C^1 solution $u = u(t, x)$ to the one-sided forward mixed problem* (1.1), (1.42) *and* (1.52) *is unique on the maximum determinate domain*

$$\{(t, x)|t \geq 0, 0 \leq x \leq x(t)\}, \tag{1.62}$$

where $x = x(t)$ is the leftmost characteristic passing through the point $(t, x) = (0, L)$:

$$\begin{cases} x'(t) = \min_{p=1,\ldots,l} \lambda_p(u(t, x(t))), \\ x(0) = L. \end{cases} \tag{1.63}$$

For getting the exact boundary controllability of nodal profile in this book, under the assumption (1.41), we should consider the following one-sided rightward mixed initial-boundary value problem for the system (1.1) with the initial data given at the t-axis:

$$x = 0: \quad u = a(t), \quad 0 \leq t \leq T_1 \tag{1.64}$$

and the boundary conditions given on the x-axis

$$t = 0: \quad v_s = l_s(\varphi(x))\varphi(x) \quad (s = m + 1, \ldots, n), \quad 0 \leq x \leq L \tag{1.65}$$

reduced from the original initial data (1.42).

Similarly to Lemma 1.4, the C^1 solution $u = u(t, x)$ to the one-sided rightward mixed initial-boundary value problem (1.1), (1.64) and (1.65) is unique on the maximum determinate domain

$$\{(t, x)|0 \leq t \leq t(x), x \geq 0\}, \tag{1.66}$$

where $t = t(x)$ denotes the downmost characteristic passing through the point $(x, t) = (0, T_1)$:

$$\begin{cases} t'(x) = \min\limits_{r=1,...,m} \dfrac{1}{\lambda_r(u(t(x), x))}, \\ t(0) = T_1. \end{cases} \tag{1.67}$$

For the solution $u = u(t, x)$ with $|u(t, x)| \leq \epsilon_0$, $\epsilon_0 > 0$ being small enough, we have always

$$t(x) \geq \underline{t}(x), \tag{1.68}$$

where $t = \underline{t}(x)$ is defined by

$$\begin{cases} \underline{t}'(x) = \inf\limits_{|u| \leq \epsilon_0} \min\limits_{r=1,...,m} \dfrac{1}{\lambda_r(u)}, \\ \underline{t}(0) = T_1, \end{cases} \tag{1.69}$$

in which

$$\beta \stackrel{\text{def.}}{=} \inf\limits_{|u| \leq \epsilon_0} \min\limits_{r=1,...,m} \dfrac{1}{\lambda_r(u)} \tag{1.70}$$

is a constant, then $t = \underline{t}(x)$ is the straight line

$$t = T_1 + \beta x. \tag{1.71}$$

Hence, the x coordinate of the intersection point of $t = \underline{t}(x)$ with the x-axis is equal to

$$-\frac{T_1}{\beta} = \frac{T_1}{\sup\limits_{|u| \leq \epsilon_0} \max\limits_{r=1,...,m} \frac{1}{|\lambda_r(u)|}}.$$

Thus, in order that the maximum determinate domain (1.66) contains the interval $[0, L]$ on the x-axis, it is sufficient to take T_1 so large that

$$T_1 \geq L \sup\limits_{|u| \leq \epsilon_0} \max\limits_{r=1,...,m} \frac{1}{|\lambda_r(u)|}. \tag{1.72}$$

In fact, in this situation, the triangular domain (see Fig. 1.1)

$$\left\{ (t, x) \Big| 0 \leq t \leq \frac{T_1}{L}(L - x), \ 0 \leq x \leq L \right\} \tag{1.73}$$

Fig. 1.1 Triangular domain
included in the maximum
determinate domain

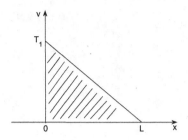

must be included in the maximum determinate domain (1.66), then the C^1 solution
$u = u(t, x)$ to the one-sided rightward mixed initial-boundary value problem (1.1),
(1.64) and (1.65) should be unique on it.

Thus, it is easy to get the following

Lemma 1.5 *Under assumption* (1.41), *for the one-sided rightward mixed initial-*
boundary value problem (1.1), (1.64) *and* (1.65), *if*

$$T_1 > L \max_{r=1,\dots,m} \frac{1}{|\lambda_r(0)|}, \tag{1.74}$$

then the maximum determinate domain of the small C^1 solution $u = u(t, x)$ with
$|u(t, x)| \le \epsilon_0$, *where $\epsilon > 0$ is small enough, must contain the interval $[0, L]$ on the*
x-axis.

For the one-sided leftward mixed initial-boundary value problem (1.1) *with the*
initial condition

$$x = L: \quad u = \bar{a}(t), \quad 0 \le t \le T_1 \tag{1.75}$$

and the boundary conditions given on the x-axis

$$t = 0: \quad v_r = l_r(\varphi(x))\varphi(x) \quad (r = 1, \dots, m), \quad 0 \le x \le L \tag{1.76}$$

reduced from the original initial data (1.42), *similar result holds, provided that* (1.74)
is replaced by

$$T_1 > L \max_{s=m+1,\dots,n} \frac{1}{\lambda_s(0)}. \tag{1.77}$$

Chapter 2
Quasilinear Wave Equations

2.1 1-D Quasilinear Wave Equations

Consider the following 1-D quasilinear wave equation

$$\frac{\partial^2 u}{\partial t^2} - \frac{\partial}{\partial x}\left(K\left(u, \frac{\partial u}{\partial x}\right)\right) = F\left(u, \frac{\partial u}{\partial x}, \frac{\partial u}{\partial t}\right), \tag{2.1}$$

where t is the time variable, x is the spatial variable, u is a scalar unknown function of (t, x), and $K = K(u, v)$ is a given C^2 function with

$$K_v(u, v) > 0, \tag{2.2}$$

and, without loss of generality, we may assume that

$$K(0, 0) = 0, \tag{2.3}$$

moreover, $F = F(u, v, w)$ is a given C^1 function, satisfying

$$F(0, 0, 0) = 0. \tag{2.4}$$

By (2.4), $u = 0$ is an equilibrium of Eq. (2.1).
   The initial condition is given as

$$t = 0: \quad u = \varphi(x), \quad u_t = \psi(x), \quad 0 \le x \le L, \tag{2.5}$$

where L is the length of the spatial interval, $\varphi(x)$ and $\psi(x)$ are C^2 and C^1 functions, respectively, on the interval $0 \le x \le L$.

© The Author(s) 2016
T. Li et al., *Exact Boundary Controllability of Nodal Profile*
for Quasilinear Hyperbolic Systems, SpringerBriefs in Mathematics,
DOI 10.1007/978-981-10-2842-7_2

On one end $x = 0$, we prescribe any one of the following boundary conditions:

$$x = 0: \quad u = h(t) \qquad \text{(Dirichlet type)}, \qquad (2.6a)$$
$$x = 0: \quad u_x = h(t) \qquad \text{(Neumann type)}, \qquad (2.6b)$$
$$x = 0: \quad u_x - \alpha u = h(t) \qquad \text{(Third type)}, \qquad (2.6c)$$
$$x = 0: \quad u_x - \beta u_t = h(t) \qquad \text{(Dissipative type)}, \qquad (2.6d)$$

where α and β are given positive constants, $h(t) \in C^2$ (for (2.6a)) or C^1 (for (2.6b)–(2.6d)) is a given function or a control function to be determined.

Similarly, on another end $x = L$, we prescribe any one of the following boundary conditions:

$$x = L: \quad u = \tilde{h}(t) \qquad \text{(Dirichlet type)}, \qquad (2.7a)$$
$$x = L: \quad u_x = \tilde{h}(t) \qquad \text{(Neumann type)}, \qquad (2.7b)$$
$$x = L: \quad u_x + \tilde{\alpha} u = \tilde{h}(t) \qquad \text{(Third type)}, \qquad (2.7c)$$
$$x = L: \quad u_x + \tilde{\beta} u_t = \tilde{h}(t) \qquad \text{(Dissipative type)}. \qquad (2.7d)$$

where $\tilde{\alpha}$ and $\tilde{\beta}$ are given positive constants, $\tilde{h}(t) \in C^2$ (for (2.7a)) or C^1 (for (2.7b)–(2.7d)) is a given function or a control function to be determined.

For the convenience of statement, in what follows we denote that

$$\delta = \begin{cases} 2 & \text{in case (2.6a)} \\ 1 & \text{in cases (2.6b)–(2.6d)} \end{cases}, \qquad \tilde{\delta} = \begin{cases} 2 & \text{in case (2.7a)} \\ 1 & \text{in cases (2.7b)–(2.7d)} \end{cases}.$$

2.2 Semi-global C^2 Solutions to the Mixed Initial-Boundary Value Problem

Similarly to Chap. 1, in order to get the exact boundary controllability of nodal profile for the quasilinear wave equation (2.1), we should first prove the existence and uniqueness of the semi-global C^2 solution to the forward mixed initial-boundary value problem of the quasilinear wave equation (2.1) with the initial condition (2.5) and with the boundary conditions (2.6)–(2.7), and the corresponding result can be presented in the following lemma (cf. [12, 19]).

Lemma 2.1 *Under the assumptions given at the beginning of Sect. 2.1, suppose furthermore that the conditions of C^2 compatibility (see Remark 2.1) are satisfied at the points $(t, x) = (0, 0)$ and $(0, L)$, respectively. Then, for any given $T_0 > 0$, the forward mixed initial-boundary value problem (2.1), (2.5) and (2.6)–(2.7) admits a unique semi-global C^2 solution $u = u(t, x)$ with small C^2 norm on the domain*

$$R(T_0) = \{(t, x) | 0 \le t \le T_0, \ 0 \le x \le L\}, \qquad (2.8)$$

provided that the norms $\|(\varphi, \psi)\|_{C^2[0,L] \times C^1[0,L]}$, $\|h\|_{C^3[0,T_0]}$ *and* $\|\tilde{h}\|_{C^3[0,T_0]}$ *are small enough (depending on T_0).*

Remark 2.1 It follows from the initial condition (2.5) that

$$u(0, 0) = \varphi(0), \quad u_x(0, 0) = \varphi'(0), \quad u_{xx}(0, 0) = \varphi''(0),$$
$$u_t(0, 0) = \psi(0), \quad u_{tx}(0, 0) = \psi'(0),$$

then from the Eq. (2.1) we have

$$u_{tt}(0, 0) = K_u(\varphi(0), \varphi'(0))\varphi'(0) + K_v(\varphi(0), \varphi'(0))\varphi''(0) + F(\varphi(0), \varphi'(0), \psi(0)).$$

Putting these values to the boundary condition (2.6) on $x = 0$ and to the differentiation of the boundary condition (2.6) on $x = 0$ with respect to t, respectively, we get the **conditions of C^2 compatibility** at the point $(t, x) = (0, 0)$.

More precisely, the conditions of C^2 compatibility at the point $(t, x) = (0, 0)$ are respectively as follows:

$$\begin{cases} \varphi(0) = h(0), \\ \psi(0) = h'(0), \\ K_u(\varphi(0), \varphi'(0))\varphi'(0) + K_v(\varphi(0), \varphi'(0))\varphi''(0) + F(\varphi(0), \varphi'(0), \psi(0)) = h''(0); \end{cases} \tag{2.9a}$$

$$\begin{cases} \varphi'(0) = h(0), \\ \psi'(0) = h'(0); \end{cases} \tag{2.9b}$$

$$\begin{cases} \varphi'(0) - \alpha\varphi(0) = h(0), \\ \psi'(0) - \alpha\psi(0) = h'(0); \end{cases} \tag{2.9c}$$

$$\begin{cases} \varphi'(0) - \beta\psi(0) = h(0), \\ \psi'(0) - \beta(K_u(\varphi(0), \varphi'(0))\varphi'(0) + K_v(\varphi(0), \varphi'(0))\varphi''(0) \\ \qquad\qquad + F(\varphi(0), \varphi'(0), \psi(0))) = h'(0). \end{cases} \tag{2.9d}$$

The conditions of C^2 compatibility at the point $(t, x) = (0, L)$ can be given in a similar way.

Proof of Lemma 2.1 Setting

$$v = \frac{\partial u}{\partial x}, \quad w = \frac{\partial u}{\partial t}, \tag{2.10}$$

the Eq. (2.1) can be reduced to the following first order quasilinear system:

$$\begin{cases} \dfrac{\partial u}{\partial t} = w, \\[2mm] \dfrac{\partial v}{\partial t} - \dfrac{\partial w}{\partial x} = 0, \\[2mm] \dfrac{\partial w}{\partial t} - K_v(u, v)\dfrac{\partial v}{\partial x} = F(u, v, w) + K_u(u, v)v \overset{\text{def.}}{=} \tilde{F}(u, v, w), \end{cases} \tag{2.11}$$

where $\tilde{F}(u, v, w)$ is still a C^1 function of u, v and w, satisfying

$$\tilde{F}(0, 0, 0) = 0. \tag{2.12}$$

It is easy to see that, (2.11) is a strictly hyperbolic system with three distinct real eigenvalues λ_i $(i = 1, 2, 3)$:

$$\lambda_1 = -\sqrt{K_v(u, v)} < \lambda_2 = 0 < \lambda_3 = \sqrt{K_v(u, v)}. \tag{2.13}$$

Thus, the characteristics for the system (2.11) are given by

$$\frac{dx_i}{dt} = \lambda_i \quad (i = 1, 2, 3). \tag{2.14}$$

Moreover, the corresponding left eigenvectors can be taken as

$$l_1 = (0, \sqrt{K_v}, 1), \quad l_2 = (1, 0, 0), \quad l_3 = (0, -\sqrt{K_v}, 1). \tag{2.15}$$

Let

$$U = (u, v, w)^T. \tag{2.16}$$

The original initial condition (2.5) can be reduced to

$$t = 0: \quad U = (\varphi(x), \varphi'(x), \psi(x))^T, \quad 0 \le x \le L, \tag{2.17}$$

and the original boundary conditions (2.6) on $x = 0$ and (2.7) on $x = L$ can be correspondingly replaced by

$$\begin{align}
x = 0: \quad & w = h'(t), & \text{(2.18a)} \\
x = 0: \quad & v = h(t), & \text{(2.18b)} \\
x = 0: \quad & v - \alpha u = h(t), & \text{(2.18c)} \\
x = 0: \quad & v - \beta w = h(t) & \text{(2.18d)}
\end{align}$$

and

$$\begin{align}
x = L: \quad & w = \tilde{h}'(t), & \text{(2.19a)} \\
x = L: \quad & v = \tilde{h}(t), & \text{(2.19b)} \\
x = L: \quad & v + \tilde{\alpha} u = \tilde{h}(t), & \text{(2.19c)} \\
x = L: \quad & v + \tilde{\beta} w = \tilde{h}(t), & \text{(2.19d)}
\end{align}$$

respectively.

Let

$$v_i = l_i(U)U \quad (i = 1, 2, 3) \tag{2.20}$$

be the diagonal variables corresponding to $\lambda_i(U)$ $(i = 1, 2, 3)$, respectively. It is easy to see that, at least in a neighbourhood of $U = 0$, the boundary conditions (2.18) and (2.19) can be rewritten as

$$x = 0: \quad v_3 = -v_1 + 2h'(t), \tag{2.21a}$$

$$x = 0: \quad v_3 = v_1 - 2\sqrt{K_v(v_2, h(t))}h(t), \tag{2.21b}$$

$$x = 0: \quad v_3 = v_1 - 2\sqrt{K_v(v_2, \alpha v_2 + h(t))}(\alpha v_2 + h(t)), \tag{2.21c}$$

$$x = 0: \quad v_3 = p_4(h(t), v_1, v_2) + \bar{p}_4(t) \tag{2.21d}$$

and

$$x = L: \quad v_1 = -v_3 + 2\tilde{h}'(t), \tag{2.22a}$$

$$x = L: \quad v_1 = v_3 + 2\sqrt{K_v(v_2, \tilde{h}(t))}\tilde{h}(t), \tag{2.22b}$$

$$x = L: \quad v_1 = v_3 + 2\sqrt{K_v(v_2, \tilde{h}(t) - \tilde{\alpha} v_2)}(\tilde{h}(t) - \tilde{\alpha} v_2), \tag{2.22c}$$

$$x = L: \quad v_1 = q_4(\tilde{h}(t), v_1, v_2) + \bar{q}_4(t), \tag{2.22d}$$

respectively.

It is easy to see that the boundary conditions (2.21) on $x = 0$ and (2.22) on $x = L$ are of the form

$$x = 0: \quad v_3 = G_3(t, v_1, v_2) + H_3(t) \tag{2.23}$$

and

$$x = L: \quad v_1 = G_1(t, v_2, v_3) + H_1(t) \tag{2.24}$$

respectively, with

$$G_1(t, 0, 0) \equiv G_3(t, 0, 0) \equiv 0. \tag{2.25}$$

Thus, Lemma 2.1 can be immediately obtained by Lemma 1.2 in the case $n = 3$.

2.3 Uniqueness of C^2 Solution to the One-Sided Mixed Initial-Boundary Value Problem

To study the exact boundary controllability of nodal profile for quasilinear wave equations, we need the uniqueness of C^2 solution to the one-sided mixed initial-boundary value problem.

For this purpose, we consider the one-sided forward mixed initial-boundary value problem for the equation (2.1) with the initial data (2.5) and the boundary conditions (2.6) on $x = 0$. By Lemma 1.4, it is easy to prove the following lemma.

Lemma 2.2 *The C^2 solution $u = u(t, x)$ to the one-sided forward mixed initial-boundary value problem (2.1) and (2.5)–(2.6) is unique on its maximum determinate domain*

$$\{(t, x)|t \geq 0, \ 0 \leq x \leq x(t)\}, \tag{2.26}$$

where $x = x(t)$ denotes the leftmost characteristic passing through the point $(t, x) = (0, L)$, i.e.,

$$\begin{cases} x'(t) = -\sqrt{K_v(u(t, x(t)), v(t, x(t)))}, \\ x(0) = L. \end{cases} \tag{2.27}$$

Noting (2.2), we now exchange the role of t and x, and consider the following one-sided rightward mixed initial-boundary value problem for the equation (2.1) with the initial data given at the t-axis:

$$x = 0: \quad u = a(t), \ u_x = b(t), \quad 0 \leq t \leq T_1, \tag{2.28}$$

and the boundary condition coming from the original initial data (2.5):

$$t = 0: \quad u = \varphi(x), \quad 0 \leq x \leq L. \tag{2.29}$$

By Lemma 2.2, the C^2 solution $u = u(t, x)$ to the one-sided rightward mixed initial-boundary value problem (2.1) and (2.28)–(2.29) is unique on its maximum determinate domain

$$\{(t, x)|0 \leq t \leq t(x), \ x \geq 0\}, \tag{2.30}$$

where $t = t(x)$ denotes the downmost characteristic passing through the point $(t, x) = (T_1, 0)$:

$$\begin{cases} t'(x) = -\dfrac{1}{\sqrt{K_v(u(t(x), x), v(t(x), x))}}, \\ t(0) = T_1. \end{cases} \tag{2.31}$$

For the solution $u = u(t, x)$ with $|u(t, x)| + |u_x(t, x)| \leq \epsilon_0, \epsilon_0 > 0$ being small enough, we have always

$$t(x) \geq \underline{t}(x), \tag{2.32}$$

where $t = \underline{t}(x)$ is defined by

$$
\begin{cases}
\underline{t}'(x) = \inf_{|u|+|v|\leq\epsilon_0} -\dfrac{1}{\sqrt{K_v(u, v)}}, \\
\underline{t}(0) = T_1,
\end{cases}
\tag{2.33}
$$

in which

$$
\beta \stackrel{\text{def.}}{=} \inf_{|u|+|v|\leq\epsilon_0} -\frac{1}{\sqrt{K_v(u, v)}}
\tag{2.34}
$$

is a constant, then $t = \underline{t}(x)$ is the straight line

$$
t = T_1 + \beta x.
\tag{2.35}
$$

Hence, the x coordinate of the intersection point of $t = \underline{t}(x)$ with the x-axis is equal to

$$
-\frac{T_1}{\beta} = \frac{T_1}{\sup\limits_{|u|+|v|\leq\epsilon_0} \frac{1}{\sqrt{K_v(u,v)}}}.
$$

Thus, in order that the maximum determinate domain (2.30) contains the interval $[0, L]$ on the x-axis, it is sufficient to take T_1 so large that

$$
T_1 \geq L \sup_{|u|+|v|\leq\epsilon_0} \frac{1}{\sqrt{K_v(u, v)}}.
\tag{2.36}
$$

In fact, in this situation, the triangular domain

$$
\left\{ (t, x) \Big| 0 \leq t \leq \frac{T_1}{L}(L - x), \ 0 \leq x \leq L \right\}
\tag{2.37}
$$

must be included in the maximum determinate domain (2.30), then the C^2 solution $u = u(t, x)$ to the one-sided rightward mixed initial-boundary value problem (2.1) and (2.28)–(2.29) should be unique on it.

Thus, we have

Lemma 2.3 *Under assumption (2.2), if*

$$
T_1 > \frac{L}{\sqrt{K_v(0, 0)}},
\tag{2.38}
$$

then for the one-sided rightward mixed initial-boundary value problem (2.1) and (2.28)–(2.29), the maximum determinate domain of the small C^2 solution $u = u(t, x)$

with $|u(t, x)| + |u_x(t, x)| \leq \epsilon_0$ ($\epsilon_0 > 0$ *being small enough) must contain the whole interval* $[0, L]$ *on the x-axis.*

For the one-sided leftward mixed initial-boundary value problem of the Eq. (2.1) with the initial condition

$$x = L: \quad u = \bar{a}(t), \ u_x = \bar{b}(t), \quad 0 \leq t \leq T_1 \tag{2.39}$$

and the boundary condition (2.29), similar results hold.

Chapter 3
Semi-global Piecewise Classical Solutions on a Tree-Like Network

3.1 Introduction

In this chapter, semi-global classical solutions on a single interval will be generalized to semi-global piecewise classical solutions on a tree-like network.

A tree-like network is a connected network without loop. Consider a tree-like network composed of N strings: $C_1, \ldots, C_N$. Without loss of generality, we suppose that one end of the string C_1 is a simple node on the network, and take this node as the starting point E.

For the ith string, let d_{i0} and d_{i1} be the x-coordinates of its two ends, and $L_i = d_{i1} - d_{i0}$ its length. For simplicity of statement, in what follows we also use d_{i0} (resp. d_{i1}) to denote the end with the coordinate d_{i0} (resp. d_{i1}). We always suppose that d_{i0} is closer to E than d_{i1} on the network. Node d_{10} is just the starting point E (see Fig. 3.1).

In what follows, let $\mathcal{M}$ and $\mathcal{S}$ be two subsets of $\{1, \ldots, N\}$, such that $i \in \mathcal{M}$ if and only if d_{i1} is a multiple node, while, $i \in \mathcal{S}$ if and only if d_{i1} is a simple node. In other words, $\mathcal{S}$ is the index set of all the simple nodes except E, and $\mathcal{M}$ is the index set of all the multiple nodes.

In Sects. 3.2 and 3.3 we will deal with semi-global piecewise classical solutions for 1-D first order quasilinear hyperbolic systems and 1-D quasilinear wave equations on a tree-like network with general topology, respectively.

3.2 Semi-global Piecewise C^1 Solutions to 1-D First Order Quasilinear Hyperbolic Systems on a Tree-Like Network

For simplicity and without loss of generality, we consider the first order quasilinear hyperbolic system of only two unknown functions with one positive eigenvalue and one negative eigenvalue, such as the Saint-Venant system for unsteady flows. In this case, we are able to use the Riemann invariants (cf. Sects. 1.3–1.4).

© The Author(s) 2016
T. Li et al., *Exact Boundary Controllability of Nodal Profile
for Quasilinear Hyperbolic Systems*, SpringerBriefs in Mathematics,
DOI 10.1007/978-981-10-2842-7_3

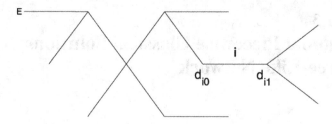

Fig. 3.1 A tree-like network

For $i = 1, \ldots, N$, we consider the following first order quasilinear hyperbolic system on the string C_i:

$$\begin{cases} \dfrac{\partial r_i}{\partial t} + \lambda_{i1}(r_i, s_i)\dfrac{\partial r_i}{\partial x} = f_{i1}(r_i, s_i), \\[2mm] \dfrac{\partial s_i}{\partial t} + \lambda_{i2}(r_i, s_i)\dfrac{\partial s_i}{\partial x} = f_{i2}(r_i, s_i), \end{cases} \qquad d_{i0} < x < d_{i1}, \qquad (3.1)$$

where (r_i, s_i) is the unknown vector function of (t, x), λ_{il} and f_{il} $(l = 1, 2)$ are C^1 functions of (r_i, s_i) with

$$f_{il}(0, 0) = 0 \quad (l = 1, 2). \qquad (3.2)$$

Suppose that on the domain under consideration we have

$$\lambda_{i1}(r_i, s_i) < 0 < \lambda_{i2}(r_i, s_i). \qquad (3.3)$$

On the simple node d_{10}, we give the nonlinear boundary condition

$$s_1 = G_{12}(t, r_1) + H_{12}(t). \qquad (3.4)$$

Similarly, on any simple node d_{i1} $(i \in \mathcal{S})$, we give the nonlinear boundary condition

$$r_i = G_{i1}(t, s_i) + H_{i1}(t). \qquad (3.5)$$

Moreover, on any multiple node d_{i1} $(i \in \mathcal{M})$, we give the nonlinear interface conditions:

$$r_i = G_{i1}(t, s_i, r_j(j \in \mathcal{J}_i)) + H_{i1}(t) \qquad (3.6)$$

and

$$s_j = G_{j2}(t, s_i, r_j(j \in \mathcal{J}_i)) + H_{j2}(t), \quad \forall j \in \mathcal{J}_i, \qquad (3.7)$$

where $\mathcal{J}_i$ denotes the set of all the indices j such that the node d_{j0} is just the node d_{i1}, namely, $\mathcal{J}_i$ is the index set of all other strings that jointly possess the multiple node d_{i1} with the ith string C_i. In (3.4)–(3.7), G_{12}, G_{i1} $(i \in \mathcal{S} \bigcup \mathcal{M})$ and G_{j2} $(j \in \mathcal{J}_i, i \in \mathcal{M})$ are C^1 functions with respect to their arguments, and, without loss of generality, we assume that

$$G_{12}(t, 0) \equiv 0, \quad G_{i1}(t, 0) \equiv 0 \quad (i \in \mathcal{S}) \tag{3.8}$$

and

$$G_{i1}(t, 0, \ldots, 0) \equiv 0 \quad (i \in \mathcal{M}), \quad G_{j2}(t, 0, \ldots, 0) \equiv 0 \quad (j \in \mathcal{J}_i, i \in \mathcal{M}). \tag{3.9}$$

The initial condition is given as follows:

$$t = 0: \quad (r_i, s_i) = (r_{i0}(x), s_{i0}(x)), \quad d_{i0} \le x \le d_{i1} \quad (i = 1, \ldots, N). \tag{3.10}$$

Noting (3.3), it is easy to see that both the boundary conditions (3.4)–(3.5) and the interface conditions (3.6)–(3.7) fit the requirements given in Sect. 1.5 for the well-posedness. That is to say, on any given simple node, the number of the boundary conditions is equal to that of the incoming characteristics, and the boundary conditions can be written in the form that the diagonal variables corresponding to these incoming characteristics are explicitly expressed by another diagonal variables; on the other hand, on any given multiple node, the number of the interface conditions is equal to that of the incoming characteristics on all the strings with this multiple node as the common node, and the interface conditions can be written in the form that the diagonal variables corresponding to these incoming characteristics are explicitly expressed by all other diagonal variables. Then, similarly to Lemma 1.1, we have

Lemma 3.1 *Under the previous assumptions, suppose furthermore that the conditions of C^1 or piecewise C^1 compatibility are satisfied on all the nodes at $t = 0$. For any given $T_0 > 0$, the forward mixed initial-boundary value problem (3.1), (3.4)–(3.7) and (3.10) admits a unique piecewise C^1 solution $(r_i(t, x), s_i(t, x))$ $(i = 1, \ldots, N)$ on the domain $R(T_0) = \bigcup_{i=1}^{N} R_i(T_0)$, where $R_i(T_0) = \{(t, x)|0 \le t \le T_0, d_{i0} \le x \le d_{i1}\}$ $(i = 1, \ldots, N)$, and the piecewise C^1 norm $\sum_{i=1}^{N} \|(r_i(\cdot, \cdot), s_i(\cdot, \cdot))\|_{C^1[R_i(T_0)]}$ is small, provided that the norms $\sum_{i=1}^{N} \|(r_{i0}(\cdot), s_{i0}(\cdot))\|_{C^1[d_{i0}, d_{i1}]}$, $\|H_{12}(\cdot)\|_{C^1[0,T_0]}$, $\|H_{i1}(\cdot)\|_{C^1[0,T_0]}$ $(i \in \mathcal{S} \bigcup \mathcal{M})$ and $\|H_{j2}(\cdot)\|_{C^1[0,T_0]}$ $(j \in \mathcal{J}_i, i \in \mathcal{M})$ are sufficiently small.*

Remark 3.1 Similar results hold for the following general 1-D quasilinear hyperbolic system:

$$\frac{\partial u_i}{\partial t} + A_i(u_i)\frac{\partial u_i}{\partial x} = F_i(u_i), \quad d_{i0} < x < d_{i1} \quad (i = 1, \ldots, N), \tag{3.11}$$

where for each $i = 1, \ldots, N$, $u_i = (u_{i1}, \ldots, u_{in})^T$ is the unknown vector function of (t, x), $A_i(u_i)$ is an $n \times n$ C^1 matrix, and $F_i(u_i) = (f_{i1}(u_i), \ldots, f_{in}(u_i))^T$ is a C^1 vector function of u_i with

$$F_i(0) = 0. \tag{3.12}$$

By hyperbolicity, for $i = 1, \ldots, N$, on the domain under consideration, the matrix $A_i(u_i)$ has n real eigenvalues $\lambda_j^{(i)}(u_i)$ $(j = 1, \ldots, n)$ and a complete set of left eigenvectors $l_j^{(i)}(u_i) = (l_{j1}^{(i)}(u_i), \ldots, l_{jn}^{(i)}(u_i))$ $(j = 1, \ldots, n)$:

$$l_j^{(i)}(u_i) A_i(u_i) = \lambda_j^{(i)}(u_i) l_j^{(i)}(u_i) \quad (j = 1, \ldots, n). \tag{3.13}$$

We assume that $\lambda_j^{(i)}(u_i)$ and $l_j^{(i)}(u_i)$ $(j = 1, \ldots, n)$ have the same C^1 regularity as $A_i(u_i)$.

For $i = 1, \ldots, N$, suppose that on the domain under consideration, we have

$$\lambda_p^{(i)}(u_i) < \lambda_q^{(i)}(u_i) \equiv 0 < \lambda_r^{(i)}(u_i)$$
$$(p = 1, \ldots, l; q = l+1, \ldots, m; r = m+1, \ldots, n). \tag{3.14}$$

Let

$$v_j^{(i)} = l_j^{(i)}(u_i) u_i \quad (j = 1, \ldots, n) \tag{3.15}$$

be the corresponding diagonal variables for $i = 1, \ldots, N$.

On the simple node d_{10}, we give the nonlinear boundary conditions

$$v_r^{(1)} = G_r^{(1)}(t, v_p^{(1)}, v_q^{(1)}) + H_r^{(1)}(t) \quad (r = m+1, \ldots, n). \tag{3.16}$$

On any simple node d_{i1} $(i \in \mathcal{S})$, we give the nonlinear boundary conditions

$$v_p^{(i)} = G_p^{(i)}(t, v_q^{(i)}, v_r^{(i)}) + H_p^{(i)}(t) \quad (p = 1, \ldots, l). \tag{3.17}$$

Moreover, on any multiple node d_{i1} $(i \in \mathcal{M})$, we give the nonlinear interface conditions

$$v_p^{(i)} = G_p^{(i)}(t, v_{l+1}^{(i)}, \ldots, v_n^{(i)}, v_1^{(j)}, \ldots, v_m^{(j)}(j \in \mathcal{J}_i)) + H_p^{(i)}(t) \quad (p = 1, \ldots, l) \tag{3.18}$$

and

$$v_r^{(j)} = G_r^{(j)}(t, v_{l+1}^{(i)}, \ldots, v_n^{(i)}, v_1^{(j)}, \ldots, v_m^{(j)}(j \in \mathcal{J}_i)) + H_r^{(j)}(t), \quad \forall j \in \mathcal{J}_i$$
$$(r = m+1, \ldots, n). \tag{3.19}$$

In (3.16)–(3.19), $G_r^{(1)}$, $G_p^{(i)}$ $(i \in \mathcal{S} \cup \mathcal{M})$ and $G_r^{(j)}$ $(j \in \mathcal{J}_i, i \in \mathcal{M})$ $(p = 1, \ldots, l; r = m+1, \ldots, n)$ are C^1 functions with respect to their arguments, and, without loss of generality, we assume that

$$G_r^{(1)}(t, 0, 0) \equiv 0 \quad (r = m+1, \ldots, n), \quad G_p^{(i)}(t, 0, 0) \equiv 0 \quad (i \in \mathcal{S}, p = 1, \ldots, l) \tag{3.20}$$

and

$$\begin{aligned} G_p^{(i)}(t, 0, \ldots, 0) &\equiv 0 \quad (i \in \mathcal{M}, p = 1, \ldots, l), \\ G_r^{(j)}(t, 0, \ldots, 0) &\equiv 0 \quad (j \in \mathcal{J}_i, i \in \mathcal{M}, r = m+1, \ldots, n). \end{aligned} \tag{3.21}$$

The initial condition is given as

$$t = 0: \quad u_i = \varphi_i(x), \quad d_{i0} \le x \le d_{i1} \quad (i = 1, \ldots, N). \tag{3.22}$$

where $\varphi_i(x) = (\varphi_{i1}(x), \ldots, \varphi_{in}(x))$ $(i = 1, \ldots, N)$.

For any given $T_0 > 0$, if the conditions of C^1 or piecewise C^1 compatibility are satisfied on all the nodes at $t = 0$, and the C^1 norms $\sum_{i=1}^{N} \|\varphi_i(\cdot)\|_{C^1[d_{i0}, d_{i1}]}$, $\|H_r^{(1)}(\cdot)\|_{C^1[0,T_0]}$ $(r = m+1, \ldots, n)$, $\|H_p^{(i)}(\cdot)\|_{C^1[0,T_0]}$ $(i \in \mathcal{S} \cup \mathcal{M}, p = 1, \ldots, l)$ and $\|H_r^{(j)}(\cdot)\|_{C^1[0,T_0]}$ $(j \in \mathcal{J}_i, i \in \mathcal{M}, r = m+1, \ldots, n)$ are sufficiently small, then the mixed initial-boundary value problem (3.11), (3.16)–(3.19) and (3.22) admits a unique piecewise C^1 solution $u_i = u_i(t, x)$ $(i = 1, \ldots, N)$ with small piecewise C^1 solution on the domain $R(T_0) = \bigcup_{i=1}^{N} R_i(T_0)$ with $R_i(T_0) = \{(t, x) | 0 \le t \le T_0, d_{i0} \le x \le d_{i1}\}$.

3.3 Semi-global Piecewise C^2 Solutions to 1-D Quasilinear Wave Equations on a Tree-Like Network

We now consider 1-D quasilinear wave equations on a tree-like network.

For $i = 1, \ldots, N$, we give the following 1-D quasilinear wave equation on the string C_i:

$$\frac{\partial^2 u_i}{\partial t^2} - \frac{\partial}{\partial x}\left(K_i\left(u_i, \frac{\partial u_i}{\partial x}\right)\right) = F_i\left(u_i, \frac{\partial u_i}{\partial x}, \frac{\partial u_i}{\partial t}\right), \tag{3.23}$$

where $K_i = K_i(u, v)$ is a given C^2 function with

$$K_{iv}(u, v) > 0, \tag{3.24}$$

and, without loss of generality, we assume that

$$K_i(0, 0) = 0; \tag{3.25}$$

moreover, $F_i = F_i(u, v, w)$ is a given C^1 function with

$$F_i(0, 0, 0) = 0. \tag{3.26}$$

The initial condition is given as

$$t = 0: \quad u_i = \varphi_i(x), \quad u_{it} = \psi_i(x), \quad d_{i0} \le x \le d_{i1} \tag{3.27}$$

for $i = 1, \ldots, N$, where $\varphi_i(x)$ and $\psi_i(x)$ are C^2 and C^1 functions, respectively.
On the simple node d_{10}, we give any one of the following boundary conditions:

$$\begin{array}{llr} u_1 = h_1(t) & \text{(Dirichlet type)}, & (3.28a) \\ u_{1x} = h_1(t) & \text{(Neumann type)}, & (3.28b) \\ u_{1x} - \alpha_1 u_1 = h_1(t) & \text{(Third type)}, & (3.28c) \\ u_{1x} - \beta_1 u_{1t} = h_1(t) & \text{(Dissipative type)}, & (3.28d) \end{array}$$

where α_1 and β_1 are given positive constants, and $h_1(t) \in C^2$ (for (3.28a)) or C^1 (for (3.28b)–(3.28d)).
On the simple node d_{i1} ($i \in \mathcal{S}$), we give any one of the following boundary conditions:

$$\begin{array}{llr} u_i = h_i(t) & \text{(Dirichlet type)}, & (3.29a) \\ u_{ix} = h_i(t) & \text{(Neumann type)}, & (3.29b) \\ u_{ix} + \alpha_i u_i = h_i(t) & \text{(Third type)}, & (3.29c) \\ u_{ix} + \beta_i u_{it} = h_i(t) & \text{(Dissipative type)}, & (3.29d) \end{array}$$

where α_i and β_i are given positive constants, and $h_i(t) \in C^2$ (for (3.29a)) or C^1 (for (3.29b)–(3.29d)).
On any multiple node d_{i1} ($i \in \mathcal{M}$), the interface conditions are given by

$$\sum_{j \in \mathcal{J}_i} K_i(u_j, u_{jx}) + K_i(u_i, u_{ix}) = h_{i0}(t) \tag{3.30}$$

and

$$u_j = u_i, \quad \forall j \in \mathcal{J}_i, \tag{3.31}$$

where $h_{i0}(t) \in C^1$ is the total stress at the multiple node d_{i1}, (3.31) means the continuity of displacement at d_{i1}, and $\mathcal{J}_i$ denotes the set of all the indices j such that the node d_{j0} is just the node d_{i1}.

By adding all the first order derivatives of u_i $(i = 1, \ldots, N)$ as new unknown variables, the mixed initial-boundary value problem (3.23) and (3.27)–(3.31) on the network can be reduced to a mixed initial-boundary value problem of the corresponding first order quasilinear hyperbolic system without zero eigenvalues. It is not difficult to check that the conclusion given in Remark 3.1 can be applied to this situation, hence we get

Lemma 3.2 *Under the previous assumptions, suppose furthermore that the conditions of C^2 or piecewise C^2 compatibility are satisfied on all the nodes at $t = 0$. Then, for any given $T_0 > 0$, the mixed initial-boundary value problem (3.23) and (3.27)–(3.31) admits a unique piecewise C^2 solution $u_i = u_i(t, x)$ $(i = 1, \ldots, N)$ with small piecewise C^2 norm on the domain $R(T_0) = \bigcup\limits_{i=1}^{N} R_i(T_0)$, where $R_i(T_0) = \{(t, x) | 0 \le t \le T_0, d_{i0} \le x \le d_{i1}\}$, provided that the norms $\sum\limits_{i=1}^{N} \|(\varphi_i(\cdot), \psi_i(\cdot))\|_{C^2[d_{i0}, d_{i1}] \times C^1[d_{i0}, d_{i1}]}, \|(h_1(\cdot), h_i(\cdot)(i \in \mathcal{S}))\|_{C^2[0, T_0]}$ (for (3.28a) and (3.29a)) or $\|(h_1(\cdot), h_i(\cdot)(i \in \mathcal{S}))\|_{C^1[0, T_0]}$ (for (3.28b)–(3.28d) and (3.29b)–(3.29d)) and $\|h_{i0}(\cdot)(i \in \mathcal{M})\|_{C^1[0, T_0]}$ are sufficiently small.*

Communications in Computer and Information Science 625

Commenced Publication in 2007
Founding and Former Series Editors:
Alfredo Cuzzocrea, Dominik Ślęzak, and Xiaokang Yang

Editorial Board

More information about this series at http://www.springer.com/series/7899

Peter Mueller · Sabu M. Thampi
Md. Zakirul Alam Bhuiyan · Ryan Ko
Robin Doss · Jose M. Alcaraz Calero (Eds.)

Security in Computing and Communications

4th International Symposium, SSCC 2016
Jaipur, India, September 21–24, 2016
Proceedings

 Springer

Editors
Peter Mueller
IBM Zurich Research Laboratory
Rueschlikon
Switzerland

Sabu M. Thampi
Technology and Management
Indian Institute of Information Technology
  and Management
Kerala
India

Md. Zakirul Alam Bhuiyan
Temple University
New York, NY
USA

Ryan Ko
Computer Science
University of Waikato
Hamilton
New Zealand

Robin Doss
Deakin University
Burwood, VIC
Australia

Jose M. Alcaraz Calero
University of the West of Scotland
Paisley, Glasgow
UK

ISSN 1865-0929   ISSN 1865-0937 (electronic)
Communications in Computer and Information Science
ISBN 978-981-10-2737-6   ISBN 978-981-10-2738-3 (eBook)
DOI 10.1007/978-981-10-2738-3

Library of Congress Control Number: 2016953330

Printed on acid-free paper

This Springer imprint is published by Springer Nature
The registered company is Springer Nature Singapore Pte Ltd.
The registered company address is: 152 Beach Road, #22-06/08 Gateway East, Singapore 189721, Singapore

Preface

The International Symposium on Security in Computing and Communications (SSCC) aims to provide the most relevant opportunity to bring together researchers and practitioners from both academia and industry to exchange their knowledge and discuss their research findings. The fourth edition, SSCC 2016 was hosted by the LNM Institute of Information Technology (LNMIIT), Jaipur (Rajasthan), India, during September 21–24, 2016. SSCC 2016 was co-located with the First International Conference on Applied Soft Computing and Communication Networks (ACN 2016).

In response to the call for papers, 136 submissions were sent to the symposium. These papers were evaluated on the basis of their significance, novelty, and technical quality. Each paper was reviewed by the members of the Program Committee and finally, 24 regular papers and 16 short papers were accepted.

There is a long list of people who volunteered their time and energy to put together the conference and who warrant acknowledgment. We would like to thank the authors of all the submitted papers, especially the accepted ones, and all the participants who made the symposium a successful event. Thanks to all members of the Technical Program Committee, and the external reviewers, for their hard work in evaluating and discussing papers.

We are grateful to the general chairs and members of the Steering Committee for their support. Our most sincere thanks go to all keynote and tutorial speakers who shared with us their expertise and knowledge. Special thanks to members of the Organizing Committee for their time and effort in organizing the conference.

We wish to express our thanks to Suvira Srivastav, Associate Editorial Director, Computer Science and Publishing Development and Publishing Editor of IAS, Springer India, for her help and cooperation. Finally, we thank Alfred Hofmann and his team at Springer for their excellent support in publishing the proceedings on time.

September 2016

Peter Mueller
Sabu M. Thampi
Md. Zakirul Alam Bhuiyan
Ryan Ko
Robin Doss
Jose M. Alcaraz Calero

Organization

Chief Patron

Lakshmi N. Mittal   LNMIIT, India
  (Chairman)

Patron

S.S. Gokhale (Director)   LNMIIT, India

General Chairs

Peter Mueller   IBM Zurich Research Laboratory, Switzerland
Sabu M. Thampi   IIITM-Kerala, India

Steering Committee

Ngoc Thanh Nguyen   Wroclaw University of Technology, Poland
Janusz Kacprzyk   Polish Academy of Sciences, Poland
Sankar Kumar Pal   Indian Statistical Institute, India
Hans-Jürgen Zimmermann   RWTH Aachen University, Germany
Nikhil R. Pal   Indian Statistical Institute, India
Sabu M. Thampi   IIITM-K, India
Mario Koeppen   Kyushu Institute of Technology, Japan
Michal Wozniak   Wroclaw University, Poland
Zoran Bojkovic   University of Belgrade, Serbia
Oge Marques   Florida Atlantic University (FAU), USA
Ranjan Gangopadhyay   LNMIIT Jaipur, India
Nabendu Chaki   University of Calcutta, India
Abdennour El Rhalibi   Liverpool John Moores University, UK
Salah Bourennane   Ecole Centrale Marseille, France
Selwyn Piramuthu   University of Florida, USA
Peter Mueller   IBM Zurich Research Laboratory, Switzerland
Robin Doss   Deakin University, Australia
Md. Zakirul Alam Bhuiyan   Temple University, USA
Axel Sikora   University of Applied Sciences Offenburg, Germany
Ryan Ko   University of Waikato, New Zealand
Sri Krishnan   Ryerson University, Toronto, Canada
El-Sayed El-Alfy   King Fahd University of Petroleum and Minerals,
           Saudi Arabia

Junichi Suzuki	University of Massachusetts Boston, USA
Parag Kulkarni	iknowlation Research Labs Pvt. Ltd., and EKLaT Research, India
Narsi Bolloju	LNMIIT Jaipur, India
Sakthi Balan	LNMIIT Jaipur, India

Organizing Chairs

Raghuvir Tomar	LNMIIT, India
Ravi Prakash Gorthi	LNMIIT, India

Secretariat

Sandeep Saini	LNMIIT, India
Kusum Lata	LNMIIT, India
Subrat Dash	LNMIIT, India

Event Management Chair

Soumitra Debnath	LNMIIT, India

Publicity Co-chair

Santosh Shah	LNMIIT, India

Program Chairs

Md. Zakirul Alam Bhuiyan	Temple University, USA
Ryan Ko	University of Waikato, New Zealand
Robin Doss	School of Information Technology, Deakin University, Australia
Jose M. Alcaraz Calero	University of the West of Scotland, UK

Program Committee/Additional Reviewers

Alireza Abdollahpouri	University of Hamburg, Germany
Mohamad Kasim Abdul Jalil	Universiti Teknologi Malaysia (UTM), Malaysia
Mohammad Faiz Liew Abdullah	Universiti Tun Hussein Onn Malaysia (UTHM), Malaysia
Davide Adami	CNIT Pisa Research Unit, University of Pisa, Italy
Rachit Adhvaryu	Gujarat Technological University, India
Sasan Adibi	Deakin University, Australia
Deepak Aeloor	Shivaji University, India

Chapter 4
Exact Boundary Controllability of Nodal Profile for 1-D First Order Quasilinear Hyperbolic Systems

4.1 Introduction

A complete theory on the local exact boundary controllability for 1-D quasilinear hyperbolic systems has been established in [11, 12, 16–18]. Recently, stimulated by some practical applications, M. Gugat, M. Herty and V. Schleper proposed in [10] a new kind of exact boundary controllability, called the nodal profile control. Different from the usual exact boundary controllability, this kind of controllability does not ask to exactly attain any given final state at a suitable time $t = T$ by means of boundary controls, instead it asks the state to exactly fit any given profile on one or some nodes after a suitable time $t = T$ by means of boundary controls. This kind of controllability, which will be certainly applicable in many practical situations, is called the exact boundary controllability of nodal profile in this book.

Based on the constructive method with modular structure suggested in [12], the exact boundary controllability of nodal profile was realized for 1-D quasilinear hyperbolic systems composed of 2 equations on certain special networks in [10]. In this Chapter, we will investigate this kind of controllability in a general framework, namely, we will give the definition of the exact boundary controllability of nodal profile and prove related results for general 1-D quasilinear hyperbolic systems with general nonlinear boundary conditions (see Li [13]). The corresponding results on a tree-like network will be given in Chap. 5.

The organization of this chapter is as follows: In Sect. 4.2 we will give the precise definitions on the exact boundary controllability of nodal profile on both boundary node and internal node, and present the corresponding results, respectively. Then, these results will be proved in Sects. 4.3 and 4.4, respectively, according to the essence of the constructive method with modular structure given in [12, 14].

© The Author(s) 2016
T. Li et al., *Exact Boundary Controllability of Nodal Profile*
for Quasilinear Hyperbolic Systems, SpringerBriefs in Mathematics,
DOI 10.1007/978-981-10-2842-7_4

4.2   Definitions and Main Results

Consider the following 1-D first order quasilinear hyperbolic system:

$$\frac{\partial u}{\partial t} + A(u)\frac{\partial u}{\partial x} = F(u), \tag{4.1}$$

where t is the time variable, x is the spatial variable, $u = (u_1, \ldots, u_n)^T$ is the unknown vector function of (t, x), $A(u)$ is a given $n \times n$ matrix with C^1 elements $a_{ij}(u)$ $(i, j = 1, \ldots, n)$, $F(u) = (f_1(u), \ldots, f_n(u))^T$ is a C^1 vector function of u and

$$F(0) = 0. \tag{4.2}$$

By (4.2), $u = 0$ is an equilibrium of system (4.1).

By hyperbolicity, for any given u on the domain under consideration, the matrix $A(u)$ possesses n real eigenvalues and a complete set of left eigenvectors $l_i(u) = (l_{i1}(u), \ldots, l_{in}(u))$ $(i = 1, \ldots, n)$:

$$l_i(u)A(u) = \lambda_i(u)l_i(u). \tag{4.3}$$

We suppose that all $\lambda_i(u)$ and $l_i(u)$ $(i = 1, \ldots, n)$ have the same C^1 regularity as $A(u) = (a_{ij}(u))$.

Suppose that on the domain under consideration there are no zero eigenvalues:

$$\lambda_r(u) < 0 < \lambda_s(u) \quad (r = 1, \ldots, m; s = m + 1, \ldots, n). \tag{4.4}$$

Let

$$v_i = l_i(u)u \quad (i = 1, \ldots, n). \tag{4.5}$$

Noting (4.4), as said in Chap. 1, for the forward problem, the most general nonlinear boundary conditions that guarantee the well-posedness can be written as

$$x = 0: \quad v_s = G_s(t, v_1, \ldots, v_m) + H_s(t) \quad (s = m + 1, \ldots, n), \tag{4.6}$$
$$x = L: \quad v_r = G_r(t, v_{m+1}, \ldots, v_n) + H_r(t) \quad (r = 1, \ldots, m), \tag{4.7}$$

where G_i and H_i $(i = 1, \ldots, n)$ are C^1 functions with respect to their arguments and, without loss of generality, we may suppose that

$$G_i(t, 0, \ldots, 0) \equiv 0 \quad (i = 1, \ldots, n). \tag{4.8}$$

Moreover, the initial condition is given as

$$t = 0: \quad u = \varphi(x), \quad 0 \leq x \leq L, \tag{4.9}$$

where L is the length of the spatial interval.

We first consider the exact boundary controllability of nodal profile on a boundary node.

Definition 4.1 (*Exact boundary controllability of nodal profile on a boundary node*)
For any given C^1 initial data $\varphi(x)$ and any given C^1 boundary functions $H_r(t)$ ($r = 1, \ldots, m$), satisfying the conditions of C^1 compatibility at the point $(t, x) = (0, L)$, for any given C^1 vector function $\bar{\bar{u}}(t)$, if there exist $T > 0$ and C^1 boundary controls $H_s(t)$ ($s = m + 1, \ldots, n$) such that the C^1 solution $u = u(t, x)$ to the mixed initial-boundary value problem (4.1), (4.6), (4.7) and (4.9) fits exactly $\bar{\bar{u}}(t)$ on $x = L$ for $t \geq T$, then we have the exact boundary controllability of nodal profile on the boundary node $x = L$.

Remark 4.1 When the exact boundary controllability of nodal profile on $x = L$ can be realized only in the case that $\varphi(x)$, $H_r(t)$ ($r = 1, \ldots, m$) and $\bar{\bar{u}}(t)$ are suitably small, it is called the local exact boundary controllability of nodal profile; otherwise, the global exact boundary controllability of nodal profile. In this book, we consider only the local exact boundary controllability of nodal profile.

Remark 4.2 In Definition 4.1, when $t \geq T$, the value of solution $u = \bar{\bar{u}}(t)$ on $x = L$ should satisfy the boundary condition (4.7), in which $v_i = \bar{\bar{v}}_i(t) \stackrel{\text{def.}}{=} l_i(\bar{\bar{u}}(t))\bar{\bar{u}}(t)$ ($i = 1, \ldots, n$). Hence, the requirement that the solution $u = u(t, x)$ fits exactly the given value $\bar{\bar{u}}(t)$ on $x = L$ for $t \geq T$ is equivalent to ask that v_s ($s = m + 1, \ldots, n$) fit exactly the given values $\bar{\bar{v}}_s(t)$ ($s = m + 1, \ldots, n$) on $x = L$ for $t \geq T$, then the value of v_r ($r = 1, \ldots, m$) on $x = L$ for $t \geq T$ can be determined by the boundary condition (4.7) as follows:

$$v_r = \bar{\bar{v}}_r(t) \stackrel{\text{def.}}{=} G_i(t, \bar{\bar{v}}_{m+1}(t), \ldots, \bar{\bar{v}}_n(t)) + H_r(t) \quad (r = 1, \ldots, m). \qquad (4.10)$$

We will prove the following theorem in Sect. 4.3.

Theorem 4.1 *Let*

$$T > L \max_{s=m+1,\ldots,n} \frac{1}{\lambda_s(0)} \qquad (4.11)$$

and let $\bar{T}$ be an arbitrarily given number such that

$$\bar{T} > T. \qquad (4.12)$$

For any given initial data $\varphi(x)$ with small C^1 norm $\|\varphi\|_{C^1[0,L]}$ and any given boundary functions $H_r(t)$ ($r = 1, \ldots, m$) with small C^1 norms $\|H_r\|_{C^1[0,\bar{T}]}$ ($r = 1, \ldots, m$), satisfying the conditions of C^1 compatibility at the point $(t, x) = (0, L)$, suppose that the given values $\bar{\bar{v}}_s(t)$ ($s = m+1, \ldots, n$) on $x = L$ for $T \leq t \leq \bar{T}$ possess small C^1 norms $\|\bar{\bar{v}}_s(t)\|_{C^1[T,\bar{T}]}$ ($s = m + 1, \ldots, n$), then there exist boundary controls $H_s(t)$ ($s = m + 1, \ldots, n$) with small C^1 norms $\|H_s\|_{C^1[0,\bar{T}]}$ ($s = m + 1, \ldots, n$), such that

the mixed initial-boundary value problem (4.1), (4.6), (4.7) *and* (4.9) *admits a unique*
C^1 *solution* $u = u(t, x)$ *with small* C^1 *norm on the domain* $R(\bar{T}) = \{(t, x)|0 \leq t \leq \bar{T}, 0 \leq x \leq L\}$, *which fits exactly the given values* $v_s = \bar{\bar{v}}_s(t)$ $(s = m + 1, \ldots, n)$, *namely, the given value* $u = \bar{\bar{u}}(t)$, *on the boundary node* $x = L$ *for* $T \leq t \leq \bar{T}$.

Remark 4.3 In Theorem 4.1, the number of boundary controls $H_s(t)$ $(s = m + 1, \ldots, n)$ is equal to that of given values $\bar{\bar{v}}_s(t)$ $(s = m + 1, \ldots, n)$.

Remark 4.4 The exact boundary controllability of nodal profile given by Definition 4.1 and discussed in Theorem 4.1 can be regarded as a kind of one-sided exact boundary controllability.

Remark 4.5 The exact boundary controllability of nodal profile on the boundary node $x = 0$ can be similarly defined and discussed. In this case, (4.11) should be replaced by

$$T > L \max_{r=1,\ldots,m} \frac{1}{|\lambda_r(0)|}. \tag{4.13}$$

We next consider the exact boundary controllability of nodal profile on an internal node.

Definition 4.2 (*Exact boundary controllability of nodal profile on an internal node*)
For any given C^1 initial data $\varphi(x)$ and any given internal node $x = \theta L$ $(0 < \theta < 1)$, for any given C^1 vector function $\bar{\bar{u}}(t)$, if there exist $T_\theta > 0$ and C^1 boundary controls $H_i(t)$ $(i = 1, \ldots, n)$ such that the C^1 solution $u = u(t, x)$ to the mixed initial-boundary value problem (4.1), (4.6), (4.7) and (4.9) fits exactly $\bar{\bar{u}}(t)$ on the internal node $x = \theta L$ for $t \geq T_\theta$, then we have the exact boundary controllability of nodal profile on the internal node $x = \theta L$.

Remark 4.6 We still have both the local exact boundary controllability of nodal profile on an internal node and the corresponding global one. In this book, we consider only the local exact boundary controllability of nodal profile on an internal node.

We will prove the following theorem in Sect. 4.4.

Theorem 4.2 *Let*

$$T_\theta > L \max_{\substack{r=1,\ldots,m \\ s=m+1,\ldots,n}} \left(\frac{1 - \theta}{|\lambda_r(0)|}, \frac{\theta}{\lambda_s(0)} \right) \tag{4.14}$$

and let $\bar{T}$ *be an arbitrarily given number such that*

$$\bar{T} > T_\theta, \tag{4.15}$$

in which $0 < \theta < 1$. *For any given initial data* $\varphi(x)$ *with small* C^1 *norm* $\|\varphi\|_{C^1[0,L]}$, *suppose that the given value* $\bar{\bar{u}}(t)$ *on* $x = \theta L$ *for* $T_\theta \leq t \leq \bar{T}$ *possesses a small* C^1 *norm* $\|\bar{\bar{u}}\|_{C^1[T_\theta,\bar{T}]}$, *then there exist boundary controls* $H_i(t)$ $(i = 1, \ldots, n)$ *with* small C^1 norms $\|H_i\|_{C^1[0,\bar{T}]}$ $(i = 1, \ldots, n)$, such that the mixed initial-boundary value problem (4.1), (4.6), (4.7) and (4.9) admits a unique C^1 solution $u = u(t, x)$ with small C^1 norm on the domain $R(\bar{T}) = \{(t, x)|0 \le t \le \bar{T}, 0 \le x \le L\}$, which fits exactly the given value $u = \bar{\bar{u}}(t)$ on the internal node $x = \theta L$ for $T_\theta \le t \le \bar{T}$.

Remark 4.7 In Theorem 4.2, the number of boundary controls $H_i(t)$ $(i = 1, \ldots, n)$ is equal to that of given values $\bar{\bar{u}}(t) = (\bar{\bar{u}}_1(t), \ldots, \bar{\bar{u}}_n(t))^T$.

Remark 4.8 The exact boundary controllability of nodal profile given by Definition 4.2 and discussed in Theorem 4.2 can be regarded as a kind of two-sided exact boundary controllability.

Remark 4.9 When $\theta \to 1$, (4.13) reduces to (4.11), and Definition 4.2 and Theorem 4.2 should be replaced by Definition 4.1 and Theorem 4.1, respectively.

Remark 4.10 It follows from (4.14) that if we take the value of θ as

$$\theta_0 = \frac{a}{a+b}, \tag{4.16}$$

in which

$$a = \max_{r=1,\ldots,m} \frac{1}{|\lambda_r(0)|}, \quad b = \max_{s=m+1,\ldots,n} \frac{1}{|\lambda_s(0)|}, \tag{4.17}$$

and consider the exact boundary controllability of nodal profile on the internal node $x = \theta_0 L$, then the optimal controllability time can be obtained as follows:

$$T_{\theta_0} > L\frac{ab}{a+b}. \tag{4.18}$$

4.3 Proof of Theorem 4.1

The constructive method with modular structure suggested in [12, 14] can be elegantly used to prove Theorem 4.1. The proof is divided into several steps.

(1) Noting (4.11), there exists an $\varepsilon_0 > 0$ so small that

$$T_1 < T, \tag{4.19}$$

where

$$T_1 = L \max_{|u| \le \varepsilon_0} \max_{s=m+1,\ldots,n} \frac{1}{\lambda_s(u)}. \tag{4.20}$$

On the domain $R(T_1) = \{(t, x)|0 \leq t \leq T_1, 0 \leq x \leq L\}$ we solve a forward mixed initial-boundary value problem for the system (4.1) with the initial condition (4.9), the boundary condition (4.7) on $x = L$ and the following artificial boundary conditions on $x = 0$:

$$x = 0: \quad v_s = g_s(t) \quad (s = m+1, \ldots, n), \tag{4.21}$$

where $g_s(t)$ $(s = m+1, \ldots, n)$ are arbitrarily given C^1 functions with small $C^1[0, T_1]$ norm, satisfying the conditions of C^1 compatibility at the point $(t, x) = (0, 0)$.

By Lemma 1.1, this forward problem admits a unique C^1 solution $u = u_f(t, x)$ with small C^1 norm on $R(T_1)$. In particular,

$$|u_f(t, x)| \leq \varepsilon_0, \quad \forall(t, x) \in R(T_1). \tag{4.22}$$

Thus, we can determine the value $\bar{u}(t)$ $(0 \leq t \leq T_1)$ of $u = u_f(t, x)$ on $x = L$ and its $C^1[0, T_1]$ norm is small.

By (4.19), there exists $u = u(t)$ on the interval $0 \leq t \leq \bar{T}$ with small $C^1[0, \bar{T}]$ norm, such that

$$u(t) = \begin{cases} \bar{u}(t), & 0 \leq t \leq T_1, \\ \bar{\bar{u}}(t), & T \leq t \leq \bar{T}, \end{cases} \tag{4.23}$$

and on the whole interval $0 \leq t \leq \bar{T}$ it satisfies the boundary condition (4.7), in which $v_i = v_i(t) \stackrel{\text{def.}}{=} l_i(u(t))u(t)$ $(i = 1, \ldots, n)$.

(2) Noting (4.4), we exchange the role of t and x, and solve a leftward mixed initial-boundary value problem on $R(\bar{T})$ for the system (4.1) with the initial condition

$$x = L: \quad u = u(t), \quad 0 \leq t \leq \bar{T}, \tag{4.24}$$

the boundary condition reduced from the original initial condition (4.9)

$$t = 0: \quad v_r = v_r(x) \stackrel{\text{def.}}{=} l_r(\varphi(x))\varphi(x) \quad (r = 1, \ldots, m), \quad 0 \leq x \leq L, \tag{4.25}$$

and the following artificial boundary conditions on $t = \bar{T}$:

$$t = \bar{T}: \quad v_s = v_s(x) \quad (s = m+1, \ldots, n), \quad 0 \leq x \leq L, \tag{4.26}$$

where $v_s(x)$ $(s = m+1, \ldots, n)$ are arbitrarily given C^1 functions with small $C^1[0, L]$ norms and satisfy the conditions of C^1 compatibility at the point $(t, x) = (\bar{T}, L)$.

According to Lemma 1.1 and Remark 1.2, this leftward problem admits a unique C^1 solution $u = u(t, x)$ with small C^1 norm on $R(\bar{T})$. In particular,

$$|u(t, x)| \leq \varepsilon_0, \quad \forall(t, x) \in R(\bar{T}). \tag{4.27}$$

(3) This C^1 solution $u = u(t, x)$ satisfies the system (4.1) and the boundary condition (4.7) on $x = L$. We now prove that $u = u(t, x)$ satisfies also the initial condition (4.9) at $t = 0$.

For this purpose, consider the following one-sided leftward mixed initial-boundary value problem for the system (4.1) with the initial condition

$$x = L: \quad u = \bar{u}(t), \quad 0 \le t \le T_1 \qquad (4.28)$$

and the boundary condition (4.25). Both $u = u(t, x)$ and $u = u_f(t, x)$ are C^1 solutions to this one-sided mixed problem. Noting (4.20), (4.22) and (4.27), by Lemma 1.5, the interval $0 \le x \le L$ on the initial axis $t = 0$ is included in the maximum determinate domain of this one-sided mixed problem, hence, by the uniqueness of C^1 solution to the one-sided mixed problem, $u = u(t, x)$ coincides with $u = u_f(t, x)$ on this interval $\{t = 0, 0 \le x \le L\}$, then $u = u(t, x)$ satisfies the initial condition (4.9) at $t = 0$.

Finally, substituting $u = u(t, x)$ into the boundary condition (4.6) on $x = 0$, we get the desired boundary controls $H_s(t)$ ($s = m+1, \ldots, n$) for $0 \le t \le \bar{T}$. This finishes the proof.

Remark 4.11 It is easily seen from the following mixed problem:

$$\frac{\partial u}{\partial t} + \frac{\partial u}{\partial x} = 0, \qquad (4.29)$$

$$x = 0: \quad u = H(t), \qquad (4.30)$$

$$t = 0: \quad u = \varphi(x), \quad 0 \le x \le L \qquad (4.31)$$

that both the estimate (4.11) on the controllability time and the number of boundary controls given by Theorem 4.1 are sharp.

4.4 Proof of Theorem 4.2

Similarly to the proof of Theorem 4.1, the proof is divided into several steps.

(1) Noting (4.14), there exists an $\varepsilon_0 > 0$ so small that

$$T_{1\theta} < T_\theta, \qquad (4.32)$$

where

$$T_{1\theta} = L \max_{|u| \le \varepsilon_0} \max_{\substack{r=1,\ldots,m \\ s=m+1,\ldots,n}} \left(\frac{1-\theta}{|\lambda_r(u)|}, \frac{\theta}{\lambda_s(u)} \right). \qquad (4.33)$$

On the domain $R(T_{1\theta}) = \{(t, x)|0 \le t \le T_{1\theta}, 0 \le x \le L\}$ we solve a forward mixed initial-boundary value problem for the system (4.1) with the initial condition (4.9), the following artificial boundary conditions on $x = 0$:

$$x = 0: \quad v_s = g_s(t) \quad (s = m + 1, \ldots, n) \tag{4.34}$$

and the following artificial boundary conditions on $x = L$:

$$x = L: \quad v_r = g_r(t) \quad (r = 1, \ldots, m), \tag{4.35}$$

where $g_i(t)$ $(i = 1, \ldots, n)$ are arbitrarily given C^1 functions with small $C^1[0, T_{1\theta}]$ norm and satisfy the conditions of C^1 compatibility at the point $(t, x) = (0, 0)$ and $(0, L)$, respectively.

By Lemma 1.1, this forward problem admits a unique C^1 solution $u = u_f(t, x)$ with small C^1 norm on $R(T_{1\theta})$. In particular,

$$|u_f(t, x)| \le \varepsilon_0, \quad \forall (t, x) \in R(T_{1\theta}). \tag{4.36}$$

Thus, we can determine the value $\bar{u}(t)$ $(0 \le t \le T_{1\theta})$ of $u = u_f(t, x)$ on $x = \theta L$ and its $C^1[0, T_{1\theta}]$ norm is small.

By (4.32), there exists $u = u(t)$ on the interval $0 \le t \le \bar{T}$ with small $C^1[0, \bar{T}]$ norm, such that

$$u(t) = \begin{cases} \bar{u}(t), & 0 \le t \le T_{1\theta}, \\ \bar{\bar{u}}(t), & T_\theta \le t \le \bar{T}. \end{cases} \tag{4.37}$$

(2) Exchanging the role of t and x, we solve a leftward (resp. rightward) mixed initial-boundary value problem on $R_l(\bar{T}) = \{(t, x)|0 \le t \le \bar{T}, 0 \le x \le \theta L\}$ (resp. $R_r(\bar{T}) = \{(t, x)|0 \le t \le \bar{T}, \theta L \le x \le L\}$) for the system (4.1) with the initial condition

$$x = \theta L: \quad u = u(t), \quad 0 \le t \le \bar{T}, \tag{4.38}$$

the boundary conditions reduced from the original initial condition (4.9)

$$t = 0: v_r = v_r(x) \stackrel{\text{def.}}{=} l_r(\varphi(x))\varphi(x) \quad (r = 1, \ldots, m), \ 0 \le x \le \theta L,$$
$$(\text{resp. } t = 0: v_s = v_s(x) \stackrel{\text{def.}}{=} l_s(\varphi(x))\varphi(x) \quad (s = m + 1, \ldots, n), \ \theta L \le x \le L), \tag{4.39}$$

and the following artificial boundary conditions on $t = \bar{T}$:

$$t = \bar{T}: v_s = \bar{v}_s(x) \quad (s = m + 1, \ldots, n), \quad 0 \le x \le \theta L,$$
$$(\text{resp. } t = \bar{T}: v_r = \bar{v}_r(x) \quad (r = 1, \ldots, m), \quad \theta L \le x \le L), \tag{4.40}$$

where $\bar{v}_i(x)$ $(i = 1, \ldots, n)$ are arbitrarily given C^1 functions with small $C^1[0, \theta L]$ or $C^1[\theta L, L]$ norms and satisfy the conditions of C^1 compatibility at the point $(t, x) = (\bar{T}, \theta L)$.

According to Lemma 1.1 and Remark 1.2, this leftward (resp. rightward) problem admits a unique C^1 solution $u = u_l(t, x)$ (resp. $u = u_r(t, x)$) on $R_l(\bar{T})$ (resp. $R_r(\bar{T})$) with small C^1 norm. In particular,

$$|u_l(t, x)| \leq \varepsilon_0, \quad \forall(t, x) \in R_l(\bar{T})$$
$$(\text{resp. } |u_r(t, x)| \leq \varepsilon_0, \quad \forall(t, x) \in R_r(\bar{T})). \tag{4.41}$$

(3) Let

$$u(t) = \begin{cases} u_l(t, x), & (t, x) \in R_l(\bar{T}), \\ u_r(t, x), & (t, x) \in R_r(\bar{T}). \end{cases} \tag{4.42}$$

$u = u(t, x)$ is a C^1 solution to the system (4.1) on the whole domain $R(\bar{T})$. We now prove that $u = u(t, x)$ satisfies the initial condition (4.9) at $t = 0$. For this purpose, consider the one-sided leftward (resp. rightward) mixed initial-boundary value problem for the system (4.1) with the initial condition

$$x = \theta L : \quad u = \bar{u}(t), \quad 0 \leq t \leq T_{1\theta} \tag{4.43}$$

and the boundary condition (4.39). Both $u = u(t, x)$ and $u = u_f(t, x)$ are C^1 solutions to this one-sided mixed problem. Noting (4.33), (4.36) and (4.41), by Lemma 1.5, the interval $0 \leq x \leq \theta L$ (resp. $\theta L \leq x \leq L$) on the initial axis $t = 0$ belongs to the corresponding maximum determinate domain, hence, by the uniqueness of C^1 solution to the one-sided mixed problem, $u = u(t, x)$ coincides with $u = u_f(t, x)$ on the interval $\{t = 0, 0 \leq x \leq L\}$, then $u = u(t, x)$ satisfies the initial condition (4.9) at $t = 0$.

Finally, substituting $u = u(t, x)$ into the boundary conditions (4.6) and (4.7), we get the desired boundary controls $H_i(t)$ $(i = 1, \ldots, n)$ for $0 \leq t \leq \bar{T}$. This ends the proof.

Remark 4.12 It is easy to see from the following mixed problem:

$$\frac{\partial u}{\partial t} - \frac{\partial u}{\partial x} = 0, \tag{4.44}$$

$$\frac{\partial v}{\partial t} + \frac{\partial v}{\partial x} = 0, \tag{4.45}$$

$$x = 0 : \quad v = H(t), \tag{4.46}$$

$$x = L : \quad u = \bar{H}(t), \tag{4.47}$$

$$t = 0 : \quad (u, v) = (\varphi(x), \psi(x)) \tag{4.48}$$

that both the estimate (4.14) on the controllability time and the number of boundary controls given by Theorem 4.2 are sharp.

4.5 Exact Boundary Controllability of Nodal Profile for Saint-Venant System on a Single Open Canal

The previous results obtained in Sects. 4.3 and 4.4 can be easily applied to the following quasilinear hyperbolic system of diagonal form:

$$
\begin{cases}
\dfrac{\partial r}{\partial t} + \lambda_1(r, s)\dfrac{\partial r}{\partial x} = f_1(r, s), \\[2mm]
\dfrac{\partial s}{\partial t} + \lambda_2(r, s)\dfrac{\partial s}{\partial x} = f_2(r, s),
\end{cases}
\quad 0 \le x \le L, \quad t \ge 0, \tag{4.49}
$$

where L is the length of the spatial variable, r and s are the unknown functions of (t, x), and $\lambda_1(r, s)$, $\lambda_2(r, s)$ and $F(r, s) = (f_1(r, s), f_2(r, s))^T$ are C^1 functions of (r, s), satisfying that on the domain under consideration

$$
\lambda_1(r, s) < 0 < \lambda_2(r, s) \tag{4.50}
$$

and

$$
F(0, 0) = 0. \tag{4.51}
$$

By (4.51), $(r, s) = (0, 0)$ is an equilibrium of system (4.49).

Give the following nonlinear boundary conditions:

$$
x = 0 : \quad s = G_2(t, r) + H_2(t), \tag{4.52}
$$
$$
x = L : \quad r = G_1(t, s) + H_1(t), \tag{4.53}
$$

where G_i and H_i $(i = 1, 2)$ are C^1 functions with respect to their arguments, and, without loss of generality, we assume that

$$
G_i(t, 0) \equiv 0 \quad (i = 1, 2). \tag{4.54}
$$

The initial condition is given as

$$
t = 0 : \quad (r, s) = (\varphi(x), \psi(x)), \quad 0 \le x \le L. \tag{4.55}
$$

By Theorem 4.1, we have

Theorem 4.3 *Let*

$$
T > \frac{L}{\lambda_2(0, 0)} \tag{4.56}
$$

and $\bar{T} > T$. For any given initial data $(\varphi(x), \psi(x))$ with small norm $\|(\varphi, \psi)\|_{C^1[0,L]}$ and for any given boundary function $H_1(t)$ with small norm

$\|H_1\|_{C^1[0,\bar{T}]}$, such that the conditions of C^1 compatibility are satisfied at the point $(t, x) = (0, L)$, if we give the nodal profile $\bar{\bar{s}}(t)$ with small norm $\|\bar{\bar{s}}\|_{C^1[T,\bar{T}]}$ at the node $x = L$ on the interval $[T, \bar{T}]$, then there exists a boundary control $H_2(t)$ with small norm $\|H_2\|_{C^1[0,\bar{T}]}$ on the interval $[0, \bar{T}]$, such that the corresponding mixed initial-boundary value problem (4.49), (4.52), (4.53) and (4.55) admits a unique C^1 solution $(r, s) = (r(t, x), s(t, x))$ with small norm $\|(r, s)\|_{C^1[R(\bar{T})]}$ on the domain $R(\bar{T}) = \{(t, x)|0 \le t \le \bar{T}, 0 \le x \le L\}$, such that at the node $x = L$ we have

$$s = \bar{\bar{s}}(t), \quad T \le t \le \bar{T}. \tag{4.57}$$

Remark 4.13 The relationship between the number of the given nodal profiles and that of the required controls given in Theorem 4.3 is as follows:

the number of the required controls $=$ the number of the given nodal profiles. (4.58)

Both sides of (4.58) are equal to 1 here. Just as pointed out in Remark 4.2, at the node $x = L$, giving the nodal profile (4.57) is equivalent to giving a pair of nodal profiles

$$\begin{cases} r = \bar{\bar{r}}(t), \\ s = \bar{\bar{s}}(t), \end{cases} \tag{4.59}$$

which satisfy the boundary condition (4.53). This means that we can give two nodal profiles (see (4.59)) at the node $x = L$, but they should satisfy one constraint (namely, the boundary condition (4.53)). Thus, (4.58) can be rewritten as

the number of the required controls
$=$ the number of the given nodal profiles $-$ the number of the constraints. (4.60)

Remark 4.14 By the proof of Theorem 4.3, we can construct a C^1 solution $(r, s) = (r(t, x), s(t, x))$ to system (4.49) on the domain $R(\bar{T})$, which satisfies the initial condition (4.55), the boundary condition (4.53) at $x = L$, and the given nodal profiles (4.59) at $x = L$ on the interval $[T, \bar{T}]$.

Remark 4.15 When the nodal profile $\bar{r}(t)$ is given at the node $x = 0$, similar result holds as in Theorem 4.3 and Remark 4.13, provided that (4.56) is replaced by

$$T > \frac{L}{|\lambda_1(0, 0)|}. \tag{4.61}$$

By Theorem 4.2, we have

Theorem 4.4 *Let*

$$T_\theta > L \max \left(\frac{1 - \theta}{|\lambda_1(0, 0)|}, \frac{\theta}{\lambda_2(0, 0)} \right) \tag{4.62}$$

and $\bar{T} > T_\theta$, *in which* $0 < \theta < 1$. *For any given initial data* $(\varphi(x), \psi(x))$ *with small norm* $\|(\varphi, \psi)\|_{C^1[0,L]}$, *if we give the nodal profile* $(\bar{r}(t), \bar{s}(t))$ *with small* C^1 *norms on the interval* $T_\theta \leq T \leq \bar{T}$ *at the internal node* $x = \theta L$, *then there exist boundary controls* $H_1(t)$ *and* $H_2(t)$ *with small* $C^1[0, \bar{T}]$ *norms, such that the corresponding mixed initial-boundary value problem* (4.49), (4.52), (4.53) *and* (4.55) *admits a unique* C^1 *solution* $(r, s) = (r(t, x), s(t, x))$ *with small* C^1 *norm on the domain* $R(\bar{T}) = \{(t, x)|0 \leq t \leq \bar{T}, 0 \leq x \leq L\}$, *such that at the internal node* $x = \theta L$ *we have*

$$r = \bar{r}(t), \quad s = \bar{s}(t), \quad T_\theta \leq t \leq \bar{T}. \tag{4.63}$$

Remark 4.16 In this case, we have (4.58), both sides of which are equal to 2, and there are no constraints, therefore (4.60) still holds.

We now consider the Saint-Venant system for unsteady flows on a single open canal.

Suppose that the canal is horizontal and cylindrical, by Sect. 1.4, the corresponding system can be written as

$$\begin{cases} \dfrac{\partial A}{\partial t} + \dfrac{\partial (AV)}{\partial x} = 0, \\[3mm] \dfrac{\partial V}{\partial t} + \dfrac{\partial S}{\partial x} = 0, \end{cases} \quad t \geq 0, \quad 0 \leq x \leq L, \tag{4.64}$$

where L is the length of the canal, $A = A(t, x)$ is the area of the cross section occupied by the water at x at time t, $V = V(t, x)$ is the average velocity over the cross section and

$$S = \frac{1}{2}V^2 + gh(A) + gY_b, \tag{4.65}$$

where g is the gravity constant, constant Y_b denotes the altitude of the bed of canal, and

$$h = h(A) \tag{4.66}$$

is the depth of the water, $h(A)$ being a suitably smooth function of A, such that

$$h'(A) > 0. \tag{4.67}$$

For system (4.64), we consider a subcritical equilibrium state (A_0, V_0) which, by definition, satisfies

$$|V_0| < \sqrt{gA_0h'(A_0)}. \tag{4.68}$$

In a neighbourhood of this equilibrium state, the eigenvalues of system (4.64) are

$$\lambda_1 \overset{\text{def.}}{=} V - \sqrt{gAh'(A)} < 0 < \lambda_2 \overset{\text{def.}}{=} V + \sqrt{gAh'(A)}. \tag{4.69}$$

The initial condition is given by

$$t = 0: \quad (A, V) = (A_0(x), V_0(x)), \quad 0 \le x \le L, \tag{4.70}$$

and the flux boundary conditions are given by

$$x = 0: \quad AV = q_2(t), \tag{4.71}$$
$$x = L: \quad AV = q_1(t). \tag{4.72}$$

Based on Theorem 4.3 we have

Theorem 4.5 *Let*

$$T > \frac{L}{\lambda_2(A_0, V_0)} \tag{4.73}$$

and $\bar{T} > T$. For any given initial data $(A_0(x), V_0(x))$ with small norm $\|(A_0(\cdot) - A_0, V_0(\cdot) - V_0)\|_{C^1[0,L]}$ and for any given boundary function $q_1(t)$ with small norm $\|q_1(\cdot) - A_0V_0\|_{C^1[0,\bar{T}]}$, such that the conditions of C^1 compatibility are satisfied at the point $(t, x) = (0, L)$, if we give the nodal profiles $(\bar{\bar{a}}(t), \bar{\bar{v}}(t))$, being the value of (A, V), at the node $x = L$ on the interval $[T, \bar{T}]$ with small norm $\|(\bar{\bar{a}}(\cdot) - A_0, \bar{\bar{v}}(\cdot) - V_0)\|_{C^1[T,\bar{T}]}$, satisfying the boundary condition (4.72), namely, $\bar{\bar{a}}(t)\bar{\bar{v}}(t) \equiv q_1(t)$ on $[T, \bar{T}]$, then there exists a boundary control $q_2(t)$ with small norm $\|q_2(\cdot) - A_0V_0\|_{C^1[0,\bar{T}]}$ on the interval $[0, \bar{T}]$, such that the corresponding mixed initial-boundary value problem (4.64), (4.70), (4.71) and (4.72) admits a unique C^1 solution $(A, V) = (A(t, x), V(t, x))$ with small norm $\|(A(\cdot, \cdot) - A_0, V(\cdot, \cdot) - V_0)\|_{C^1[R(\bar{T})]}$ on the domain $R(\bar{T}) = \{(t, x)|0 \le t \le \bar{T}, 0 \le x \le L\}$. Moreover, at the boundary node $x = L$ the solution exactly satisfies the given nodal profiles

$$A = \bar{\bar{a}}(t), \quad V = \bar{\bar{v}}(t), \quad T \le t \le \bar{T}. \tag{4.74}$$

Remark 4.17 When the nodal profiles are given at the node $x = 0$, similar result holds, provided that (4.71) is replaced by

$$T > \frac{L}{|\lambda_1(A_0, V_0)|}. \tag{4.75}$$

Proof of Theorem 4.4 First, we reduce system (4.62) to a system of diagonal form.

For this purpose, we introduce the Riemann invariants (r, s) (cf. Sect. 1.4):

$$\begin{cases} 2r = V - V_0 - G(A), \\ 2s = V - V_0 + G(A), \end{cases} \tag{4.76}$$

where

$$G(A) = \int_{A_0}^A \sqrt{\frac{gh'(A)}{A}} \, dA. \tag{4.77}$$

Then

$$\begin{cases} V = r + s + V_0, \\ A = H(s - r) > 0, \end{cases} \tag{4.78}$$

where H is the inverse function of G and

$$H(0) = A_0, \tag{4.79}$$

$$H'(0) = \sqrt{\frac{A_0}{gh'(A_0)}} > 0. \tag{4.80}$$

$(A, V) = (A_0, V_0)$ is equivalent to $(r, s) = (0, 0)$.

In a neighbourhood of (A_0, V_0), system (4.64) can be equivalently rewritten as

$$\begin{cases} \dfrac{\partial r}{\partial t} + \lambda_1(r, s)\dfrac{\partial r}{\partial x} = 0, \\[2mm] \dfrac{\partial s}{\partial t} + \lambda_2(r, s)\dfrac{\partial s}{\partial x} = 0, \end{cases} \quad t \geq 0, \quad 0 \leq x \leq L, \tag{4.81}$$

where

$$\begin{cases} \lambda_1(r, s) = r + s + V_0 - \sqrt{gH(s - r)h'(H(s - r))} < 0, \\ \lambda_2(r, s) = r + s + V_0 + \sqrt{gH(s - r)h'(H(s - r))} > 0. \end{cases} \tag{4.82}$$

Correspondingly, the boundary conditions (4.71) and (4.72) can be rewritten as

$$x = 0: \quad P_2 \stackrel{\text{def.}}{=} (r + s + V_0)H(s - r) - q_2(t) = 0, \tag{4.83}$$

$$x = L: \quad P_1 \stackrel{\text{def.}}{=} (r + s + V_0)H(s - r) - q_1(t) = 0. \tag{4.84}$$

Since, when $(r, s) = (0, 0)$,

$$\frac{\partial P_2}{\partial s} = \sqrt{\frac{A_0}{gh'(A_0)}} \left(\sqrt{gA_0 h'(A_0)} + V_0 \right) > 0, \tag{4.85}$$

$$\frac{\partial P_1}{\partial r} = \sqrt{\frac{A_0}{gh'(A_0)}} \left(\sqrt{gA_0 h'(A_0)} - V_0 \right) < 0, \tag{4.86}$$

the boundary conditions (4.83) and (4.84) can be equivalently rewritten as

$$x = 0 : \quad s = G_2(t, r) + H_2(t), \tag{4.87}$$

$$x = L : \quad r = G_1(t, s) + H_1(t), \tag{4.88}$$

where, without loss of generality, we assume that

$$G_1(t, 0) \equiv G_2(t, 0) \equiv 0, \tag{4.89}$$

then

$$\|q_i(\cdot) - A_0 V_0\|_{C^1[0, \bar{T}]} \text{ small} \iff \|H_i(\cdot)\|_{C^1[0, \bar{T}]} \text{ small} \quad (i = 1, 2). \tag{4.90}$$

Moreover, at $x = L$ on the interval $[T, \bar{T}]$, giving the nodal profiles (4.74) is equivalent to giving the nodal profiles

$$r = \frac{1}{2} \left(\bar{v}(t) - V_0 - G(\bar{a}(t)) \right), \quad s = \frac{1}{2} \left(\bar{v}(t) - V_0 + G(\bar{a}(t)) \right). \tag{4.91}$$

Thus, by Theorem 4.3 and Remark 4.14, we can construct a C^1 solution $(A, V) = (A(t, x), V(t, x))$ to system (4.64) on the domain $R(\bar{T})$, which satisfies the initial condition (4.70), the boundary condition (4.72) at $x = L$ and the nodal profiles (4.74) at $x = L$ on the interval $[T, \bar{T}]$. This finishes the proof of Theorem 4.4.

Remark 4.18 We use Fig. 4.1 to illustrate the exact boundary controllability of nodal profile on a single open canal given by Theorem 4.5. In this figure, "→" on one node means giving a nodal profile or a pair of nodal profiles with constraint on this node, while "•" on one node means choosing a control on this node.

Similarly, by Theorem 4.4 we have

Theorem 4.6 *Let*

Fig. 4.1 Exact boundary controllability of nodal profile at a boundary node on a single open canal

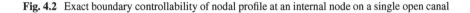

Fig. 4.2 Exact boundary controllability of nodal profile at an internal node on a single open canal

$$T_\theta > L \max \left(\frac{1-\theta}{|\lambda_1(A_0, V_0)|}, \frac{\theta}{\lambda_2(A_0, V_0)} \right) \tag{4.92}$$

and $\bar{T} > T_\theta$, in which $0 < \theta < 1$. For any given initial data $(A_0(x), V_0(x))$ with small norm $\|(A_0(\cdot) - A_0, V_0(\cdot) - V_0)\|_{C^1[0,L]}$, if we give the nodal profile $(\tilde{\bar{a}}(t), \tilde{\bar{v}}(t))$ at the internal node $x = \theta L$ on the interval $[T_\theta, \bar{T}]$ with the small norm $\|(\tilde{\bar{a}}(\cdot) - A_0, \tilde{\bar{v}}(\cdot) - V_0)\|_{C^1[T_\theta, \bar{T}]}$, then there exist boundary controls $q_1(t)$ and $q_2(t)$ with small norm $\|q_i(\cdot) - A_0 V_0\|_{C^1[0,\bar{T}]}$ $(i = 1, 2)$, such that the corresponding mixed initial-boundary value problem (4.64), (4.70), (4.71) and (4.72) admits a unique C^1 solution $(A, V) = (A(t, x), V(t, x))$ with small norm $\|(A(\cdot, \cdot) - A_0, V(\cdot, \cdot) - V_0)\|_{C^1[R(\bar{T})]}$ on the domain $R(\bar{T}) = \{(t, x)|0 \le t \le \bar{T}, 0 \le x \le L\}$, such that at the internal node $x = \theta L$ the solution exactly satisfies the given nodal profiles (4.74).

Remark 4.19 Using the notations given in Remark 4.18, the exact boundary controllability of nodal profile at an internal node can be illustrated by Fig. 4.2.

Remark 4.20 Theorems 4.5 and 4.6 are still valid if the flux boundary conditions (4.71) and (4.72) are replaced by

$$x = 0: \quad S = s_2(t), \tag{4.93}$$
$$x = L: \quad S = s_1(t), \tag{4.94}$$

where $\|s_i(\cdot) - S_0\|_{C^1[0,\bar{T}]}$ is small enough, in which $S_0 = \frac{1}{2}V_0^2 + gh(A_0) + gY_b$.

4.6   Remarks

Like in Sect. 4.5 for Saint-Venant system, the exact boundary controllability of nodal profile for other relevant physical models without zero eigenvalues can be also realized, for example, see [2].

The whole discussion in this chapter is for the case that there are no zero eigenvalues (see (4.4)). When there are zero eigenvalues such that (1.51) holds, in order to get the exact controllability of nodal profile, as in the case of the exact controllability, in general we have to use not only boundary controls but also internal controls. For this, we refer the readers to the recent results given in [23, 24].

Chapter 5
Exact Boundary Controllability of Nodal Profile for 1-D First Order Quasilinear Hyperbolic Systems on a Tree-Like Network

In this chapter we will generalize the exact boundary controllability of nodal profile for 1-D first order quasilinear hyperbolic systems on a single spatial interval, discussed in Chap. 4, to that on a tree-like network. A general framework can be established for general 1-D first order quasilinear hyperbolic systems with general nonlinear boundary conditions and general nonlinear interface conditions, provided that there are full of boundary controls in both boundary conditions and interface conditions (see Gu and Li [8]).

However, from the point of view of practical applications, we still face on some serious problems. First, both the nodal profiles and the controls are given in the form of diagonal variables in [8], but in practical problems, the nodal profiles are given in a form of physically meaningful functions, and the controls are usually physical variables which can be chosen easily, however, the nodal profiles and the controls have often quite complicated relationships with the diagonal variables of the system. Second, the boundary controls are usually not enough in practical problems, namely, in certain boundary conditions or interface conditions, we can not find enough boundary controls. These lead new challenges for establishing the exact boundary controllability of nodal profile on a tree-like network.

In this chapter, facing these difficulties and taking the Saint-Venant system as example, we study the exact boundary controllability of nodal profile for unsteady flows on a tree-like network of open canals, in which the nodal profiles are given in the form of the area A of the cross section occupied by the water (or the depth h of the water) and the average velocity V of the water, while the controls are chosen to be the flux on the corresponding nodes. Based on it, we get a general theory on the exact boundary controllability of nodal profile for unsteady flows on a tree-like network of open canals, and give the possibility to establish the controllability of nodal profile by only choosing the controls on simple nodes (see Gu and Li [9]. For the corresponding result on the exact boundary controllability, cf. Gu and Li [7]).

© The Author(s) 2016

T. Li et al., *Exact Boundary Controllability of Nodal Profile*
for Quasilinear Hyperbolic Systems, SpringerBriefs in Mathematics,
DOI 10.1007/978-981-10-2842-7_5

5.1   Exact Boundary Controllability of Nodal Profile for Saint-Venant System on a Star-Like Network of Open Canals (Case 1)

In this section, we first consider the exact boundary controllability of nodal profile for Saint-Venant system on a star-like network of open canals.

A star-like network is a connected network with only one multiple node. Suppose that the network is composed of N open canals with the joint point O (see Fig. 5.1). For $i = 1, \cdots, N$, let E_i be another node of the ith canal, and L_i its length. Let the coordinate of the multiple node O be $x = 0$, the ith canal can be parameterized as $x \in [0, L_i]$ $(i = 1, \cdots, N)$.

Still suppose that the canals are horizontal and cylindrical, similarly to the case of one single canal (see Sect. 1.4), the unsteady flows on a star-like network of open canals can be described by the following Saint-Venant system:

$$\begin{cases} \dfrac{\partial A_i}{\partial t} + \dfrac{\partial (A_i V_i)}{\partial x} = 0, \\[2mm] \dfrac{\partial V_i}{\partial t} + \dfrac{\partial S_i}{\partial x} = 0, \end{cases} \quad t \geq 0, \quad 0 \leq x \leq L_i \ (i = 1, \cdots, N), \qquad (5.1)$$

where, for the ith canal, $A_i = A_i(t, x)$ is the area of the cross section occupied by the water at x at time t, $V_i = V_i(t, x)$ is the average velocity over the cross section and

$$S_i = \frac{1}{2} V_i^2 + g h_i(A_i) + g Y_{ib}, \qquad (5.2)$$

where g is the gravity constant, constant Y_{ib} denotes the altitude of the bed of canal and

$$h_i = h_i(A_i) \qquad (5.3)$$

Fig. 5.1 Star-like network of open canals

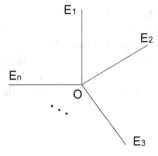

is the depth of the water, $h_i(A_i)$ being a suitably smooth function of A_i, such that

$$h_i'(A_i) > 0. \tag{5.4}$$

The initial condition is given as follows:

$$t = 0: \quad (A_i, V_i) = (A_{i0}(x), V_{i0}(x)), \quad 0 \le x \le L_i \quad (i = 1, \cdots, N). \tag{5.5}$$

At any given simple node we have the flux boundary condition

$$x = L_i: \quad A_i V_i = q_i(t) \quad (i = 1, \cdots, N), \tag{5.6}$$

while, at the multiple node O we have the total flux interface condition

$$x = 0: \quad \sum_{i=1}^{N} A_i V_i = q_0(t), \tag{5.7}$$

and the "energy-type" interface condition

$$x = 0: \quad S_i = S_1 \quad (i = 2, \cdots, N). \tag{5.8}$$

Consider subcritical equilibrium states $(A_i, V_i) = (A_{i0}, V_{i0})$ $(i = 1, \cdots, N)$ of system (5.1). By definition, they satisfy

$$|V_{i0}| < \sqrt{gA_{i0}h_i'(A_{i0})} \quad (i = 1, \cdots, N). \tag{5.9}$$

Moreover, corresponding to (5.7)–(5.8), we suppose that

$$\sum_{i=1}^{N} A_{i0} V_{i0} = 0, \tag{5.10}$$

$$S_{i0} = S_{10} \quad (i = 2, \cdots, N), \tag{5.11}$$

where

$$S_{i0} = \frac{1}{2} V_{i0}^2 + gh_i(A_{i0}) + gY_{bi} \quad (i = 1, \cdots, N). \tag{5.12}$$

In a neighbourhood of this equilibrium state, the eigenvalues of system (5.1) are

$$\lambda_{i1} \overset{\text{def.}}{=} V_i - \sqrt{gA_i h_i'(A_i)} < 0 < \lambda_{i2} \overset{\text{def.}}{=} V_i + \sqrt{gA_i h_i'(A_i)} \quad (i = 1, \cdots, N). \tag{5.13}$$

We first consider the case that the nodal profiles are given at a simple node (say $x = L_1$) and the boundary control is taken at the multiple node $x = 0$ (**case 1**). We have

Theorem 5.1 *Let*

$$T > \frac{L_1}{\lambda_{12}(A_{10}, V_{10})} \tag{5.14}$$

and $\bar{T} > T$. For any given initial data $(A_{i0}(x), V_{i0}(x))$ $(i = 1, \cdots, N)$ with small norm $\sum_{i=1}^{N} \|(A_{i0}(\cdot) - A_{i0}, V_{i0}(\cdot) - V_{i0})\|_{C^1[0,L_i]}$ and for any given boundary flux functions $q_i(t)$ $(i = 1, \cdots, N)$ with small norm $\|q_i(\cdot) - A_{i0}V_{i0}\|_{C^1[0,\bar{T}]}$ $(i = 1, \cdots, N)$, such that the conditions of C^1 compatibility are satisfied at the points $(t, x) = (0, L_i)$ $(i = 1, \cdots, N)$, respectively, if we give the nodal profiles $(\bar{a}_1(t), \bar{v}_1(t))$, being the value of (A_1, V_1), at the simple node $x = L_1$ on the interval $[T, \bar{T}]$ with small norm $\|(\bar{a}_1(\cdot) - A_{10}, \bar{v}_1(\cdot) - V_{10})\|_{C^1[T,\bar{T}]}$, satisfying the flux boundary condition (5.6) at $x = L_1$, namely, $\bar{a}_1(t)\bar{v}_1(t) \equiv q_1(t)$ for $T \le t \le \bar{T}$, then there exists a boundary control $q_0(t)$ with small norm $\|q_0(\cdot)\|_{C^1[0,\bar{T}]}$ at the multiple node $x = 0$ on the interval $[0, \bar{T}]$, such that the corresponding mixed initial-boundary value problem (5.1) and (5.5)–(5.8) admits a unique piecewise C^1 solution $(A_i(t, x), V_i(t, x))$ $(i = 1, \cdots, N)$ with small norm $\sum_{i=1}^{N} \|(A_i(\cdot, \cdot) - A_{i0}, V_i(\cdot, \cdot) - V_{i0})\|_{C^1[R_i(\bar{T})]}$ on the domain $R(\bar{T}) = \bigcup_{i=1}^{N} R_i(\bar{T}) = \bigcup_{i=1}^{N}\{(t, x)|0 \le t \le \bar{T}, 0 \le x \le L_i\}$, such that at the boundary node $x = L_1$ the solution exactly satisfies the given nodal profiles

$$A_1 = \bar{a}_1(t), \quad V_1 = \bar{v}_1(t), \quad T \le t \le \bar{T}. \tag{5.15}$$

Proof By Lemma 3.1, under the assumptions of Theorem 5.1, it is easy to check that for any given $q_i(t)$ $(i = 0, 1, \cdots, N)$, the mixed initial-boundary value problem (5.1) and (5.5)–(5.8) admits a unique piecewise C^1 solution $(A_i(t, x), V_i(t, x))$ $(i = 1, \cdots, N)$. Hence, to prove this theorem, we need only to construct a piecewise C^1 solution to system (5.1) on the domain $R(\bar{T})$, such that it satisfies the initial condition (5.5), the boundary condition (5.6) at $x = L_i$ $(i = 1, \cdots, N)$, the "energy-type" interface condition (5.8) at $x = 0$ and the given nodal profiles (5.15) at $x = L_1$ on the interval $[T, \bar{T}]$. Substituting this solution into the total flux interface condition (5.7) at $x = 0$, we get the desired control $q_0(t)$.

For this end, we first consider the first canal. Noting (5.14), by Theorem 4.3 and Remark 4.14, on the domain $R_1(\bar{T})$ we can find a C^1 solution $(A_1(t, x), V_1(t, x))$ to system (5.1) for $i = 1$, such that it satisfies the initial condition (5.5) for $i = 1$, the boundary condition (5.6) at $x = L_1$, and the nodal profiles (5.15) at $x = L_1$ on the interval $[T, \bar{T}]$. Thus, we can get the value of S_1 at $x = 0$:

$$S_1(t) = \frac{1}{2}V_1^2(t, 0) + gh_1(A_1(t, 0)) + gY_{1b}, \quad 0 \le t \le \bar{T}. \tag{5.16}$$

Next, for each $k = 2, \cdots, N$, we want to find a C^1 solution $(A_k(t, x), V_k(t, x))$ to system (5.1) for $i = k$ on the domain $R_k(\bar{T})$, such that it satisfies the initial condition (5.5) for $i = k$, the boundary condition (5.6) at $x = L_k$ and the boundary condition

$$x = 0: \quad S_k = S_1(t). \tag{5.17}$$

For this purpose, similarly to the proof of Theorem 4.3, we introduce Riemann invariants (r_k, s_k):

$$\begin{cases} 2r_k = V_k - V_{k0} - G_k(A_k), \\ 2s_k = V_k - V_{k0} + G_k(A_k), \end{cases} \tag{5.18}$$

where

$$G_k(A_k) = \int_{A_{k0}}^{A_k} \sqrt{\frac{gh'_k(A_k)}{A_k}} \, dA_k. \tag{5.19}$$

Then,

$$\begin{cases} V_k = r_k + s_k + V_{k0}, \\ A_k = H_k(s_k - r_k) > 0, \end{cases} \tag{5.20}$$

where H_k is the inverse function of G_k with

$$H_k(0) = A_{k0}, \tag{5.21}$$

$$H'_k(0) = \sqrt{\frac{A_{k0}}{gh'_k(A_{k0})}} > 0. \tag{5.22}$$

Thus, the kth system in (5.1) can be equivalently rewritten as

$$\begin{cases} \dfrac{\partial r_k}{\partial t} + \lambda_{k1}(r_k, s_k)\dfrac{\partial r_k}{\partial x} = 0, \\ \dfrac{\partial s_k}{\partial t} + \lambda_{k2}(r_k, s_k)\dfrac{\partial s_k}{\partial x} = 0, \end{cases} \quad t \geq 0, \quad 0 \leq x \leq L_k, \tag{5.23}$$

where

$$\begin{cases} \lambda_{k1}(r_k, s_k) = r_k + s_k + V_{k0} - \sqrt{gH_k(s_k - r_k)h'_k(H_k(s_k - r_k))} < 0, \\ \lambda_{k2}(r_k, s_k) = r_k + s_k + V_{k0} + \sqrt{gH_k(s_k - r_k)h'_k(H_k(s_k - r_k))} > 0. \end{cases} \tag{5.24}$$

The boundary condition (5.6) at $x = L_k$ can be equivalently rewritten as

$$r_k = G_{k1}(t, s_k) + H_{k1}(t), \tag{5.25}$$

where, without loss of generality, we assume that

$$G_{k1}(t, 0) \equiv 0, \tag{5.26}$$

then

$$\|q_k(\cdot) - A_{k0}V_{k0}\|_{C^1[0,\bar{T}]} \text{ small} \iff \|H_{k1}(\cdot)\|_{C^1[0,\bar{T}]} \text{ small.} \tag{5.27}$$

Let

$$P_{3k} \stackrel{\text{def.}}{=} S_k(r_k, s_k) - S_1(t). \tag{5.28}$$

Noting that, when $(A_k, V_k) = (A_{k0}, V_{k0})$, we have

$$\frac{\partial P_{3k}}{\partial s_k} = \sqrt{gA_{k0}h'_k(A_{k0})} + V_{k0} > 0, \tag{5.29}$$

then, in a neighbourhood of $(A_k, V_k) = (A_{k0}, V_{k0})$, the boundary condition (5.17) can be equivalently rewritten as

$$x = 0: \quad s_k = G_{k2}(t, r_k) + H_{k2}(t), \tag{5.30}$$

where, without loss of generality, we assume that

$$G_{k2}(t, 0) \equiv 0, \tag{5.31}$$

then $\|h_{k2}(\cdot)\|_{C^1[0,\bar{T}]}$ is suitably small. By Lemma 1.1, the mixed initial-boundary value problem of system (5.1) with the initial condition (5.5) for $i = k$, the boundary condition (5.6) at $x = L_k$ and the boundary condition (5.17) at $x = 0$ admit a unique C^1 solution $(A_k(t, x), V_k(t, x))$ with small C^1 norm on the domain $R_k(\bar{T})$ corresponding to the kth canal. Thus, $(A_i(t, x), V_i(t, x))$ $(i = 1, \cdots, N)$ is a desired piecewise C^1 solution to system (5.1) on the whole domain $R(\bar{T})$.

Remark 5.1 In Theorem 5.1, the number of the given nodal profiles and that of the boundary controls still satisfy the relationship 4.60. In this situation, 2 nodal profiles are given at the simple node L_1, but they should satisfy one boundary condition on this node. The number of the corresponding boundary controls is 1, and this control is given at another node of the corresponding canal, namely, the multiple node O. See Fig. 5.2.

Remark 5.2 In Theorem 5.1, at any given single node $x = L_i$ of a star-like network, the flux boundary condition (5.6) can be replaced by

$$x = L_i: \quad S_i = s_i(t), \tag{5.32}$$

Fig. 5.2 Exact boundary
controllability of nodal
profile on a star-like network
(case 1)

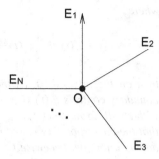

where $s_i(t)$ is a given C^1 function with small norm $\|s_i(\cdot) - S_{i0}\|_{C^1[0,\tilde{T}]}$. (5.32) is still
a well-posed boundary condition.

Remark 5.3 When the total flux $q_0(t)$ at the multiple node $x = 0$ is taken as
the boundary control, the nodal profiles can be given only on one simple node
(cf. (4.60)).

5.2   Exact Boundary Controllability of Nodal Profile for Saint-Venant System on a Star-Like Network of Open Canals (Case 2)

We now consider the case that the nodal profiles are given at the multiple node $x = 0$
for some canals, while the boundary controls are taken at the multiple node $x = 0$
and at the simple nodes of these canals (**case 2**). We have

Theorem 5.2 *Let* $1 \le m < N$,

$$T > \max_{i=1,\cdots,m} \frac{L_i}{|\lambda_{i1}(A_{i0}, V_{i0})|} \tag{5.33}$$

and $\tilde{T} > T$. *For any given initial data* $(A_{i0}(x), V_{i0}(x))$ $(i = 1, \cdots, N)$ *with small
norm* $\sum_{i=1}^{N} \|(A_{i0}(\cdot) - A_{i0}, V_{i0}(\cdot) - V_{i0})\|_{C^1[0,L_i]}$ *and for any given boundary flux
functions* $q_i(t)$ $(i = m + 1, \cdots, N)$ *with small norms* $\|q_i(\cdot) - A_{i0}V_{i0}\|_{C^1[0,\tilde{T}]}$ $(i = m+1, \cdots, N)$, *such that the conditions of* C^1 *compatibility are satisfied at the points
$(t, x) = (0, L_i)$ $(i = m + 1, \cdots, N)$, *respectively, if for* $i = 1, \cdots, m$, *we give the
nodal profiles* $(\bar{a}_i(t), \bar{v}_i(t))$, *being the value of* (A_i, V_i), *at* $x = 0$ *on the interval
$[T, \tilde{T}]$ with small norms* $\|(\bar{a}_i(\cdot) - A_{i0}, \bar{v}_i(\cdot) - V_{i0})\|_{C^1[T,\tilde{T}]}$ *on the ith canal, satisfying
the "energy-type" interface condition*

$$x = 0: \quad \bar{\bar{S}}_i(t) = \bar{\bar{S}}_1(t) \quad (i = 2, \cdots, m), \tag{5.34}$$

where

$$\bar{\bar{S}}_i(t) = \frac{1}{2}\bar{\bar{v}}_i(t)^2 + gh_i(\bar{\bar{a}}_i(t)) + gY_{ib} \quad (i = 1, \cdots, m), \tag{5.35}$$

then on the interval $[0, \bar{T}]$ *there exists a boundary control* $q_0(t)$ *at* $x = 0$ *and boundary controls* $q_i(t)$ $(i = 1, \cdots, m)$ *at* $x = L_i$ $(i = 1, \cdots, m)$, *respectively, with small norms* $\|q_0(\cdot)\|_{C^1[0,\bar{T}]}$ *and* $\|q_i(\cdot) - A_{i0}V_{i0}\|_{C^1[0,\bar{T}]}$ $(i = 1, \cdots, m)$, *such that the corresponding mixed initial-boundary value problem* (5.1) *and* (5.5)–(5.8) *admits a unique piecewise* C^1 *solution* $(A_i(t, x), V_i(t, x))$ $(i = 1, \cdots, N)$ *on the domain* $R(\bar{T})$ *with small norm* $\sum_{i=1}^N \|(A_i(\cdot, \cdot) - A_{i0}, V_i(\cdot, \cdot) - V_{i0}\|_{C^1[R_i(\bar{T})]}$, *such that at* $x = 0$ *the solution satisfies the given nodal profiles*

$$A_i = \bar{\bar{a}}_i(t), \quad V_i = \bar{\bar{v}}_i(t) \quad (i = 1, \cdots, m), \quad T \le t \le \bar{T}. \tag{5.36}$$

Proof As in the proof of Theorem 5.1, we need only to construct a piecewise C^1 solution to system (5.1) on the domain $R(\bar{T})$, such that it satisfies the initial condition (5.5), the boundary conditions (5.6) at $x = L_i$ $(i = m + 1, \cdots, N)$, the "energy-type" interface condition (5.8) at $x = 0$ and the given nodal profiles (5.36) at $x = 0$ on the interval $[T, \bar{T}]$.

First, introducing an artificial flux boundary condition at $x = 0$ on the first canal, such that it is consistent with the given nodal profiles (5.36) for $i = 1$. Noting (5.33), by Theorem 4.3 and Remark 4.14, we can construct a C^1 solution $(A_1(t, x), V_1(t, x))$ to system (5.1) for $i = 1$ on the domain $R_1(\bar{T})$, which satisfies the initial condition (5.5) for $i = 1$ and the nodal profiles (5.36) for $i = 1$ at $x = 0$ on the interval $[T, \bar{T}]$. Then, we can get the value of S_1 at $x = 0$:

$$S_1(t) = \frac{1}{2}V_1^2(t, 0) + gh_1(A_1(t, 0)) + gY_{1b}, \quad 0 \le t \le \bar{T}. \tag{5.37}$$

For each $k = 2, \cdots, m$, by Remark 5.2, the condition $S_k = S_1(t)$ can be treated as a well-posed boundary condition at $x = 0$ on the kth canal, moreover, the given nodal profiles $(\bar{\bar{a}}_k(t), \bar{\bar{v}}_k(t))$ satisfy this constraint. Hence, for $k = 2, \cdots, m$, noting (5.33), on the kth canal we can also construct a C^1 solution $(A_k(t, x), V_k(t, x))$ to system (5.1) for $i = k$ on the domain $R_k(\bar{T})$, which satisfies the initial condition (5.5) for $i = k$, the "energy-type" interface condition (5.8) for $i = k$ at $x = 0$ and the given nodal profile (5.36) for $i = k$ at $x = 0$ on the interval $[T, \bar{T}]$.

Finally, for each $k = m + 1, \cdots, N$, we can solve system (5.1) for $i = k$ with the initial condition (5.5) for $i = k$, the flux boundary condition (5.6) at $x = L_k$ and the boundary condition (5.8) at $x = 0$ on the domain $R_k(\bar{T})$. Noting Remark 5.2, by Lemma 1.1, this mixed initial-boundary value problem admits a unique C^1 solution $(A_k(t, x), V_k(t, x))$. Thus, $(A_i(t, x), V_i(t, x))$ $(i = 1, \cdots, N)$ is a desired piecewise C^1 solution to system (5.1) on the entire domain $R(\bar{T})$.

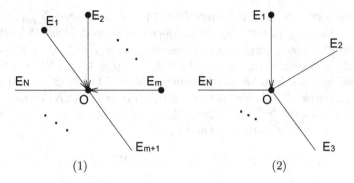

Fig. 5.3 Exact boundary controllability of nodal profile on a star-like network (case 2)

Remark 5.4 In Theorem 5.2, the number of the given nodal profiles and that of the boundary controls still satisfy the relationship 4.60. In this situation, there are $2m$ nodal profiles given at the multiple node $x = 0$, but they should satisfy $(m - 1)$ constraints given by the "energy-type" interface conditions at this node. The number of the corresponding boundary controls is $m + 1$, in which one is given at the multiple node $x = 0$, while the other m boundary controls are given at another nodes of the canals corresponding to the given nodal profiles, respectively. See Fig. 5.3(1), where "•" still means choosing a control at this node, while "→" means giving a pair of nodal profiles corresponding to that canal at this node. When $m = 1$, there are 2 given nodal profiles without constraints, and 2 controls are chosen at two nodes of the first canal, respectively. See Fig. 5.3(2).

5.3   Exact Boundary Controllability of Nodal Profile for Saint-Venant System on a Star-Like Network of Open Canals (Case 3)

Corresponding to the missing case $m = N$ in Theorem 5.2, we now consider the case that the nodal profiles are given at the multiple node $x = 0$ for all the canals, while the boundary controls are taken at all the simple nodes (**case 3**). We have

Theorem 5.3 *Let*

$$T > \max_{i=1,\cdots,N} \frac{L_i}{|\lambda_{i1}(A_{i0}, V_{i0})|} \tag{5.38}$$

and $\bar{T} > T$. *For any given initial data* $(A_{i0}(x), V_{i0}(x))$ $(i = 1, \cdots, N)$ *with small norm* $\sum_{i=1}^{N} \|(A_{i0}(\cdot) - A_{i0}, V_{i0}(\cdot) - V_{i0})\|_{C^1[0, L_i]}$ *and for any given total flux* $q_0(t)$ *with small norm* $\|q_0(\cdot)\|_{C^1[0, \bar{T}]}$ *such that the conditions of piecewise* C^1 *compatibility are satisfied at the point* $(t, x) = (0, 0)$, *if we give the nodal profiles* $(\bar{a}_i(t), \bar{v}_i(t))$ $(i = 1, \cdots, N)$, *being the values of* (A_i, V_i) $(i = 1, \cdots, N)$, *respectively, for* all the canals at $x = 0$ on the interval $[T, \bar{T}]$ with small norm $\sum_{i=1}^{N} \|(\bar{\bar{a}}_i(\cdot) - A_{i0}, \bar{\bar{v}}_i(\cdot) - V_{i0})\|_{C^1[T,\bar{T}]}$, satisfying the interface conditions (5.7) and (5.8), then on the interval $[0, \bar{T}]$ there exist boundary controls $q_i(t)$ at $x = L_i$ $(i = 1, \cdots, N)$, respectively, with small norms $\|q_i(\cdot) - A_{i0}V_{i0}\|_{C^1[0,\bar{T}]}$ $(i = 1, \cdots, N)$, such that the corresponding mixed initial-boundary value problem (5.1) and (5.5)–(5.8) admits a unique piecewise C^1 solution $(A_i(t, x), V_i(t, x))$ $(i = 1, \cdots, N)$ on the domain $R(\bar{T})$ with small norm $\sum_{i=1}^{N} \|(A_i(\cdot, \cdot) - A_{i0}, V_i(\cdot, \cdot) - V_{i0})\|_{C^1[R_i(\bar{T})]}$, such that at $x = 0$ the solution satisfies the given nodal profiles

$$A_i = \bar{\bar{a}}_i(t), \quad V_i = \bar{\bar{v}}_i(t) \quad (i = 1, \cdots, N), \quad T \le t \le \bar{T}. \qquad (5.39)$$

Proof As in the proof of Theorem 5.1, we need only to construct a piecewise C^1 solution $(A_i(t, x), V_i(t, x))$ $(i = 1, \cdots, N)$ to system (5.1) on the domain $R(\bar{T})$, such that it satisfies the initial condition (5.5), the interface conditions (5.7) and (5.8), and the given nodal profiles (5.39) at $x = 0$ on the interval $[T, \bar{T}]$.

Noting (5.38), there exists an $\varepsilon_0 > 0$ so small that

$$T_1 \stackrel{\text{def.}}{=} \max_{i=1,\cdots,N} \max_{|(A_i-A_{i0}, V_i-V_{i0})| \le \varepsilon_0} \frac{L_i}{|\lambda_{i1}(A_i, V_i)|} < T. \qquad (5.40)$$

Giving artificial boundary conditions at $x = L_i$:

$$x = L_i : \quad A_i V_i = Q_i(t) \quad (i = 1, \cdots, N), \qquad (5.41)$$

where $Q_i(t)$ $(i = 1, \cdots, N)$ are suitably smooth functions with small norm $\sum_{i=1}^{N} \|Q_i(\cdot) - A_{i0}V_{i0}\|_{C^1[0,T_1]}$ and satisfy the corresponding conditions of C^1 compatibility at $x = L_i$ $(i = 1, \cdots, N)$, respectively, by Lemma 3.1, the mixed initial-boundary value problem (5.1), (5.5), (5.41) and (5.7)–(5.8) admits a unique piecewise C^1 solution $(\bar{A}_i(t, x), \bar{V}_i(t, x))$ $(i = 1, \cdots, N)$ on the domain $R(T_1)$ with small norm $\sum_{i=1}^{N} \|(\bar{A}_i(\cdot, \cdot) - A_{i0}, \bar{V}_i(\cdot, \cdot) - V_{i0})\|_{C^1[R_i(T_1)]}$. Thus, we get $(\bar{a}_i(t), \bar{v}_i(t))$, the value of $(\bar{A}_i(t, x), \bar{V}_i(t, x))$ at $x = 0$, on the interval $[0, T_1]$ for all $i = 1, \cdots, N$. Since $T_1 < T$, we can get smooth functions $(a_i(t), v_i(t))$ $(i = 1, \cdots, N)$ at $x = 0$ with small C^1 norms $\|(a_i(\cdot) - A_{i0}, v_i(\cdot) - V_{i0})\|_{C^1[0,\bar{T}]}$ $(i = 1, \cdots, N)$, such that for $i = 1, \cdots, N$,

$$(a_i(t), v_i(t)) = \begin{cases} (\bar{a}_i(t), \bar{v}_i(t)), & 0 \le t \le T_1, \\ (\bar{\bar{a}}_i(t), \bar{\bar{v}}_i(t)), & T \le t \le \bar{T}, \end{cases} \qquad (5.42)$$

and $(a_i(t), v_i(t))$ $(i = 1, \cdots, N)$ satisfy the interface conditions (5.7)–(5.8) at $x = 0$ on the whole interval $[0, \bar{T}]$.

Next, exchanging the role of t and x, for each $i = 1, \cdots, N$, we solve a rightward mixed initial-boundary value problem for system (5.1) on the domain $R_i(\bar{T})$ together with the initial condition

$$x = 0 : \quad (A_i, V_i) = (a_i(t), v_i(t)), \quad 0 \le t \le \bar{T}, \qquad (5.43)$$

the boundary condition reduced from the original initial condition (5.5):

$$t = 0: \quad A_i V_i = A_{i0}(x) V_{i0}(x), \quad 0 \leq x \leq L_i \tag{5.44}$$

and a suitable artificial boundary condition

$$t = \bar{T}: \quad A_i V_i = Q_i(x), \quad 0 \leq x \leq L_i. \tag{5.45}$$

By Lemma 1.1, these rightward mixed problems admit uniquely C^1 solutions $(A_i(t, x), V_i(t, x))$ with small C^1 norms of $(A_i(t, x) - A_{i0}, V_i(t, x) - V_{i0})$ on $R_i(\bar{T})$ for $i = 1, \cdots, N$, respectively.

Obviously, for each $i = 1, \cdots, N$, both $(A_i(t, x), V_i(t, x))$ and $(\bar{A}_i(t, x), \bar{V}_i(t, x))$ satisfy system (5.1) on $R_i(T_1)$, the initial condition

$$x = 0: \quad (A_i, V_i) = (\bar{a}_i(t), \bar{v}_i(t)), \quad 0 \leq t \leq T_1 \tag{5.46}$$

and the boundary condition (5.44). Noting (5.38), by Lemma 1.5, it is easy to see that for $i = 1, \cdots, N$ we have

$$(A_i(0, x), V_i(0, x)) = (\bar{A}_i(0, x), \bar{V}_i(0, x)) = (A_{i0}(x), V_{i0}(x)), \quad 0 \leq x \leq L_i, \tag{5.47}$$

then $(A_i(t, x), V_i(t, x))$ $(i = 1, \cdots, N)$, as a piecewise C^1 solution to system (5.1), satisfies the initial condition (5.5), the interface conditions (5.7)–(5.8), and the given nodal profiles (5.39) at $x = 0$. This complete the proof.

Remark 5.5 In Theorem 5.3, the number of the given nodal profiles and that of the boundary controls still satisfy the relationship (4.60). In fact, in this case, there are $2N$ nodal profiles given at the multiple node $x = 0$, but they should satisfy N constraints, namely, the interface conditions (5.7) and (5.8). Moreover, the boundary controls are flux functions at all the simple nodes, the number of which is N. See Fig. 5.4, where "•" still means choosing a control at this node, while "→" means giving a pair of node profiles on the corresponding canal at this node.

Fig. 5.4 Exact boundary controllability of nodal profile on a star-like network (case 3)

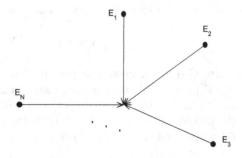

5.4   Exact Boundary Controllability of Nodal Profile for Saint-Venant System on a Star-Like Network of Open Canals (Case 4)

We now consider the case that the nodal profiles are given at one simple node (say $x = L_1$), while the boundary control is taken at another simple node (say $x = L_N$) (**case 4**). We have

Theorem 5.4 *Let*

$$T > \frac{L_1}{\lambda_{12}(A_{10}, V_{10})} + \frac{L_N}{|\lambda_{N1}(A_{N0}, V_{N0})|} \tag{5.48}$$

and $\bar{T} > T$. For any given initial data $(A_{i0}(x), V_{i0}(x))$ $(i = 1, \cdots, N)$ with small norm $\sum_{i=1}^{N} \|(A_{i0}(\cdot) - A_{i0}, V_{i0}(\cdot) - V_{i0})\|_{C^1[0,L_i]}$ and for any given functions $q_0(t)$ and $q_i(t)$ $(i = 1, \cdots, N-1)$ with small norms $\|q_0(\cdot)\|_{C^1[0,\bar{T}]}$ and $\|q_i(\cdot) - A_{i0}V_{i0}\|_{C^1[0,\bar{T}]}$ $(i = 1, \cdots, N-1)$, such that the conditions of C^1 compatibility are satisfied at the points $(t, x) = (0, 0)$ and $(0, L_i)$ $(i = 1, \cdots, N-1)$, respectively, if we give the nodal profiles $(\bar{\bar{a}}_1(t), \bar{\bar{v}}_1(t))$, being the value of (A_1, V_1), at $x = L_1$ on the interval $[T, \bar{T}]$ with small norm $\|(\bar{\bar{a}}_1(\cdot) - A_{10}, \bar{\bar{v}}_1(\cdot) - V_{10})\|_{C^1[T,\bar{T}]}$, satisfying the flux boundary condition at $x = L_1$, namely, $\bar{\bar{a}}_1(t)\bar{\bar{v}}_1(t) \equiv q_1(t)$ for $T \leq t \leq \bar{T}$, then there exists a boundary control $q_N(t)$ at $x = L_N$ on the interval $[0, \bar{T}]$ with small norm $\|q_N(\cdot) - A_{N0}V_{N0}\|_{C^1[0,\bar{T}]}$, such that the mixed initial-boundary value problem (5.1) and (5.5)–(5.8) admits a unique piecewise C^1 solution $(A_i(t, x), V_i(t, x))$ $(i = 1, \cdots, N)$ on the domain $R(\bar{T})$ with small norm $\sum_{i=1}^{N} \|(A_i(\cdot, \cdot) - A_{i0}, V_i(\cdot, \cdot) - V_{i0})\|_{C^1[R_i(\bar{T})]}$, such that at $x = L_1$ the solution satisfies the nodal profiles

$$A_1 = \bar{\bar{a}}_1(t), \quad V_1 = \bar{\bar{v}}_1(t), \quad T \leq t \leq \bar{T}. \tag{5.49}$$

Proof Noting (5.48), there exists an $\varepsilon_0 > 0$ so small that

$$T_1 \stackrel{\text{def.}}{=} \max_{|(A_1 - A_{10}, V_1 - V_{10})| \leq \varepsilon_0} \frac{L_1}{\lambda_{12}(A_1, V_1)} + \max_{|(A_N - A_{N0}, V_N - V_{N0})| \leq \varepsilon_0} \frac{L_N}{|\lambda_{N1}(A_N, V_N)|}$$
$$< T. \tag{5.50}$$

Giving an artificial boundary condition at $x = L_N$:

$$x = L_N : \quad A_N V_N = Q_N(t), \tag{5.51}$$

where $Q_N(t)$ is a suitably smooth function such that the C^1 norm of $(Q_N(t) - A_{N0}V_{N0})$ is suitably small and the corresponding conditions of C^1 compatibility are satisfied at $x = L_N$, by Lemma 3.1, the mixed initial-boundary value problem (5.1), (5.5), (5.6) for $i = 1, \cdots, N - 1$, (5.7)–(5.8) and (5.51) admits a unique piecewise C^1 solution $(\bar{A}_i(t, x), \bar{V}_i(t, x))$ $(i = 1, \cdots, N)$ on the domain $R(T_1)$ with small norm $\sum_{i=1}^{N} \|(\bar{A}_i(\cdot, \cdot) - A_{i0}, \bar{V}_i(\cdot, \cdot) - V_{i0})\|_{C^1[R_i(T_1)]}$. Thus, we get $(\bar{a}_1(t), \bar{v}_1(t))$, the value of $(\bar{A}_1(t, x), \bar{V}_1(t, x))$ at $x = L_1$ on the interval $[0, T_1]$. Since $T_1 < T$, we can get a smooth function $(a_1(t), v_1(t))$ with small $C^1[0, \bar{T}]$ norm at $x = L_1$, such that

$$(a_1(t), v_1(t)) = \begin{cases} (\bar{a}_1(t), \bar{v}_1(t)), & 0 \le t \le T_1, \\ (\bar{\bar{a}}_1(t), \bar{\bar{v}}_1(t)), & T \le t \le \bar{T}, \end{cases} \tag{5.52}$$

and $(a_1(t), v_1(t))$ satisfies the boundary condition (5.6) for $i = 1$ at $x = L_1$ on the interval $[0, \bar{T}]$.

Next, exchanging the role of t and x, we solve a leftward mixed initial-boundary value problem for system (5.1) for $i = 1$ on the domain $R_1(\bar{T})$ together with the initial condition

$$x = L_1: \quad (A_1, V_1) = (a_1(t), v_1(t)), \quad 0 \le t \le \bar{T}, \tag{5.53}$$

the boundary condition reduced from the original initial condition (5.5) for $i = 1$:

$$t = 0: \quad A_1 V_1 = A_{10}(x) V_{10}(x), \quad 0 \le x \le L_1 \tag{5.54}$$

and an artificial boundary condition

$$t = \bar{T}: \quad A_1 V_1 = Q_1(x), \quad 0 \le x \le L_1, \tag{5.55}$$

where $Q_1(x)$ is a suitably smooth function with small norm $\|Q_1(\cdot) - A_{10} V_{10}\|_{C^1[0,L_1]}$, satisfying the conditions of C^1 compatibility at the point $(t, x) = (\bar{T}, L)$. By Lemma 1.1, this leftward mixed problem admits a unique C^1 solution $(A_1(t, x), V_1(t, x))$ with small C^1 norm of $(A_1(t, x) - A_{10}, V_1(t, x) - V_{10})$ on the domain $R_1(\bar{T})$.

Obviously, both $(A_1(t, x), V_1(t, x))$ and $(\bar{A}_1(t, x), \bar{V}_1(t, x))$ satisfy system (5.1) for $i = 1$ on the domain $R_1(T_1)$, the initial condition

$$x = L_1: \quad (A_1, V_1) = (\bar{a}_1(t), \bar{v}_1(t)), \quad 0 \le t \le T_1 \tag{5.56}$$

and the boundary condition (5.54). Noting (5.48), by Lemma 1.5 it is easy to see that

$$(A_1(0, x), V_1(0, x)) = (\bar{A}_1(0, x), \bar{V}_1(0, x)) = (A_{10}(x), V_{10}(x)), \quad 0 \le x \le L_1 \tag{5.57}$$

and

$$(A_1(t, 0), V_1(t, 0)) = (\bar{A}_1(t, 0), \bar{V}_1(t, 0)), \quad 0 \le t \le T_2, \tag{5.58}$$

where

$$T_2 = \max_{|(A_N - A_{N0}, V_N - V_{N0})| \le \varepsilon_0} \frac{L_N}{|\lambda_{N1}(A_N, V_N)|}. \tag{5.59}$$

Let

$$S_1(t) = \frac{1}{2} V_1^2(t, 0) + g h_1(A_1(t, 0)) + g Y_{1b}, \quad 0 \le t \le \bar{T}. \tag{5.60}$$

For each $k = 2, \cdots, N - 1$, on the kth canal we solve a forward mixed initial-boundary value problem for system (5.1) for $i = k$ on the domain $R_k(\bar{T})$ with the initial condition (5.5) for $i = k$, the boundary condition (5.6) at $x = L_k$ and the boundary condition

$$x = 0: \quad S_k = S_1(t). \tag{5.61}$$

Noting Remark 5.2, this problem admits a unique C^1 solution $(A_k(t, x), V_k(t, x))$ on the domain $R_k(\bar{T})$ with small C^1 norm of $(A_k(t, x) - A_{k0}, V_k(t, x) - V_{k0})$. Moreover, noting that both this solution and $(\bar{A}_k(t, x), \bar{V}_k(t, x))$ satisfy the same system with the same initial and boundary conditions on the domain $R_k(T_2)$, noting (5.48) and (5.59), by Lemma 1.5, at $x = 0$ we have

$$(A_k(t, 0), V_k(t, 0)) = (\bar{A}_k(t, 0), \bar{V}_k(t, 0)), \quad 0 \le t \le T_2 \quad (k = 2, \cdots, N - 1). \tag{5.62}$$

Let

$$P_4 \overset{\text{def.}}{=} A_N V_N + \sum_{i=1}^{N-1} A_i V_i - q_0(t) \tag{5.63}$$

and

$$P_5 \overset{\text{def.}}{=} S_N(A_N, V_N) - S_1(t). \tag{5.64}$$

Noting that, when $(A_N, V_N) = (A_{N0}, V_{N0})$, we have

$$\left| \frac{\partial(P_4, P_5)}{\partial(A_N, V_N)} \right| = V_{N0}^2 - g A_{N0} h'_N(A_{N0}) < 0, \tag{5.65}$$

we get that, in a neighbourhood of $(A_N, V_N) = (A_{N0}, V_{N0})$, the total flux interface condition and the "energy-type" interface condition for $i = N$ at $x = 0$ can be equivalently rewritten in a form that A_N and V_N can be expressed by (A_k, V_k) ($i = 1, \cdots, N - 1$). Since we have already constructed the solution $(A_i(t, x), V_i(t, x))$ on the domain $R_i(\bar{T})$ for $i = 1, \cdots, N - 1$, we can use the value of these solutions at $x = 0$ on the interval $[0, \bar{T}]$ to uniquely determine $(a_N(t), v_N(t))$, the value of

(A_N, V_N) at $x = 0$ on the interval $[0, \bar{T}]$, moreover, by (5.58) and (5.62) we have

$$(a_N(t), v_N(t)) = (A_N(t, 0), V_N(t, 0)) = (\bar{A}_N(t, 0), \bar{V}_N(t, 0)), \quad 0 \le t \le T_2.$$
$$(5.66)$$

Finally, in order to construct a C^1 solution on the Nth canal, we solve a rightward mixed initial-boundary value problem for system (5.1) for $i = N$ on the domain $R_N(\bar{T})$ with the initial condition

$$x = 0: \quad (A_N, V_N) = (a_N(t), v_N(t)), \quad 0 \le t \le \bar{T}, \tag{5.67}$$

the boundary condition reduced by the original initial condition (5.5):

$$t = 0: \quad A_N V_N = A_{N0}(x) V_{N0}(x), \quad 0 \le x \le L_N \tag{5.68}$$

and an artificial boundary condition

$$t = \bar{T}: \quad A_N V_N = \tilde{Q}_N(x), \quad 0 \le x \le L_N, \tag{5.69}$$

where $\tilde{Q}_N(t)$ is a suitably smooth function with small norm $\|\tilde{Q}_N(\cdot) - A_{N0} V_{N0}\|_{C^1[0, L_N]}$. By Lemma 1.1, this rightward mixed problem admits a unique C^1 solution $(A_N(t, x), V_N(t, x))$ with small C^1 norm of $(A_N(t, x) - A_{N0}, V_N(t, x) - V_{N0})$ on the domain $R_N(\bar{T})$.

Noting (5.66), both $(A_N(t, x), V_N(t, x))$ and $(\bar{A}_N(t, x), \bar{V}_N(t, x))$ satisfy system (5.1) for $i = N$ on the domain $R_N(T_2)$, the initial condition

$$x = 0: \quad (A_N, V_N) = (a_N(t), v_N(t)), \quad 0 \le t \le T_2 \tag{5.70}$$

and the boundary condition (5.68). Noting (5.48) and (5.59), by Lemma 1.5, we get

$$(A_N(0, x), V_N(0, x)) = (\bar{A}_N(0, x), \bar{V}_N(0, x)) = (A_{N0}(x), V_{N0}(x)), \quad 0 \le x \le L_N.$$
$$(5.71)$$

Thus, we find a piecewise C^1 solution $(A_i(t, x), V_i(t, x))$ $(i = 1, \cdots, N)$ to system (5.1) on the domain $R(\bar{T})$, which satisfies the initial condition (5.5), the boundary conditions (5.6) at $x = L_i$ $(i = 1, \cdots, N - 1)$, the interface conditions (5.7)–(5.8) at $x = 0$, and the nodal profiles given at $x = L_1$ on the interval $[T, \bar{T}]$. This finishes the proof.

Remark 5.6 In Theorem 5.4, the number of the given nodal profiles and that of the boundary controls still satisfy the relationship (4.60). In this situation, 2 nodal profiles are given at the simple node $x = L_1$, but they should satisfy one constraint given by the boundary condition at this node, while there is 1 boundary control given at one of other simple nodes, for example, at $x = L_N$. See Fig. 5.5.

Fig. 5.5 Exact boundary
controllability of nodal
profile on a star-like network
(case 4)

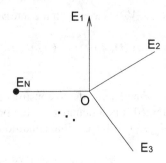

5.5 Exact Boundary Controllability of Nodal Profile for Saint-Venant System on a Star-Like Network of Open Canals (Case 5)

Finally, we consider the case that the nodal profiles are given at the multiple node $x = 0$ on one of the canals (say, on the first canal), while the boundary controls are taken at both the simple node of the corresponding canal (say, on $x = L_1$) and one of other simple nodes (say $x = L_N$) (**case 5**). In this situation, it is not necessary to ask this pair of nodal profiles given at $x = 0$ to have a constraint.

Similarly to Theorem 5.4, we have

Theorem 5.5 *Let*

$$T > \max\left(\frac{L_1}{|\lambda_{11}(A_{10}, V_{10})|}, \frac{L_N}{|\lambda_{N1}(A_{N0}, V_{N0})|}\right) \tag{5.72}$$

and $\bar{T} > T$. For any given initial data $(A_{i0}(x), V_{i0}(x))$ $(i = 1, \cdots, N)$ with small norm $\sum_{i=1}^{N} \|(A_{i0}(\cdot) - A_{i0}, V_{i0}(\cdot) - V_{i0}\|_{C^1[0, L_i]}$ and for any given functions $q_0(t)$ and $q_i(t)$ $(i = 2, \cdots, N-1)$ with small norms $\|q_0(\cdot)\|_{C^1[0, \bar{T}]}$ and $\|q_i(\cdot) - A_{i0}V_{i0}\|_{C^1[0, \bar{T}]}$ $(i = 2, \cdots, N-1)$, such that the conditions of C^1 compatibility are satisfied at the points $(t, x) = (0, 0)$ and $(0, L_i)$ $(i = 2, \cdots, N-1)$, respectively, if we give the nodal profiles $(\bar{a}_1(t), \bar{v}_1(t))$, being the value of (A_1, V_1), at $x = 0$ on the interval $[T, \bar{T}]$ with small norm $\|(\bar{a}_1(\cdot) - A_{10}, \bar{v}_1(\cdot) - V_{10}\|_{C^1[T, \bar{T}]}$ on the first canal, then there exist boundary controls $q_1(t)$ at $x = L_1$ and $q_N(t)$ at $x = L_N$ on the interval $[0, \bar{T}]$ with small norms $\|q_i(\cdot) - A_{i0}V_{i0}\|_{C^1[0, \bar{T}]}$ $(i = 1, N)$, such that the corresponding mixed initial-boundary value problem (5.1) and (5.5)–(5.8) admits a unique piecewise C^1 solution $(A_i(t, x), V_i(t, x))$ $(i = 1, \cdots, N)$ on the domain $R(\bar{T})$ with small norm $\sum_{i=1}^{N} \|(A_i(\cdot, \cdot) - A_{i0}, V_i(\cdot, \cdot) - V_{i0}\|_{C^1[R_i(\bar{T})]}$, such that at the multiple node $x = 0$ the solution satisfies the given nodal profiles

$$A_1 = \bar{a}_1(t), \quad V_1 = \bar{v}_1(t), \quad T \le t \le \bar{T}. \tag{5.73}$$

Proof By (5.72), there exists an $\varepsilon_0 > 0$ so small that

$$T_1 \overset{\text{def.}}{=} \max \left(\max_{|(A_1 - A_{10}, V_1 - V_{10})| \leq \varepsilon_0} \frac{L_1}{|\lambda_{11}(A_1, V_1)|}, \right.$$
$$\left. \max_{|(A_N - A_{N0}, V_N - V_{N0})| \leq \varepsilon_0} \frac{L_N}{|\lambda_{N1}(A_N, V_N)|} \right) < T. \qquad (5.74)$$

Introducing suitable artificial boundary conditions at $x = L_1$ and L_N, respectively. By Lemma 3.1, we can get a piecewise C^1 solution $(\bar{A}_i(t, x), \bar{V}_i(t, x))$ $(i = 1, \cdots, N)$ to system (5.1) on the domain $R(T_1)$ with small piecewise C^1 norm of $(\bar{A}_i(t, x) - A_{i0}, \bar{V}_i(t, x) - V_{i0})$ $(i = 1, \cdots, N)$, and this solution satisfies the initial condition (5.5), the interface conditions (5.7)–(5.8) and the boundary conditions (5.6) for $i = 2, \cdots, N - 1$. Hence, we can get $(\bar{a}_1(t), \bar{v}_1(t))$, the value of $(\bar{A}_1(t, x), \bar{V}_1(t, x))$ at $x = 0$ on the interval $[0, T_1]$. Since $T_1 < T$, we can construct a smooth function $(a_1(t), v_1(t))$ at $x = 0$ with small $C^1[0, \bar{T}]$ norm, such that

$$(a_1(t), v_1(t)) = \begin{cases} (\bar{a}_1(t), \bar{v}_1(t)), & 0 \leq t \leq T_1, \\ (\bar{\bar{a}}_1(t), \bar{\bar{v}}_1(t)), & T \leq t \leq \bar{T}. \end{cases} \qquad (5.75)$$

Then, similarly to the proof of Theorem 5.4, with the initial condition $(A_1, V_1) = (a_1(t), v_1(t))$ $(0 \leq t \leq \bar{T})$ at $x = 0$, the boundary condition reduced from the original initial condition (5.5) for $i = 1$

$$t = 0: \quad A_1 V_1 = A_{10}(x) V_{10}(x), \quad 0 \leq x \leq L_1, \qquad (5.76)$$

and an artificial boundary condition given at $t = \bar{T}$, we solve a rightward mixed initial-boundary value problem for system (5.1) for $i = 1$ on the first canal to get a C^1 solution $(A_1(t, x), V_1(t, x))$ on the domain $R_1(\bar{T})$ with small C^1 norm of $(A_1(t, x) - A_{10}, V_1(t, x) - V_{10})$, and, noting (5.74), by Lemma 1.5 this solution satisfies the initial condition (5.5) for $i = 1$.

Next, for each $k = 2, \cdots, N-1$, on the kth canal we solve a forward mixed initial-boundary value problem on the domain $R_k(\bar{T})$ to get a C^1 solution $(A_k(t, x), V_k(t, x))$ to system (5.1) for $i = k$ with the initial condition (5.5), the boundary condition (5.6) and the "energy-type" interface condition (5.8) all for $i = k$.

Finally, by the total flux interface condition (5.7) and the "energy-type" interface condition (5.8) for $i = N$, we can uniquely determine the value of $(A_N(t, x), V_N(t, x))$ at $x = 0$. Taking this value as the initial data at $x = 0$, together with the boundary condition reduced from the original initial condition (5.5) for $i = N$

$$t = 0: \quad A_N V_N = A_{N0}(x) V_{N0}(x), \quad 0 \leq x \leq L_N, \qquad (5.77)$$

and an artificial boundary condition given at $t = \bar{T}$ on the Nth canal, we solve a rightward mixed initial-boundary value problem for system (5.1) for $i = N$ to get

Fig. 5.6 Exact boundary
controllability of nodal
profile on a star-like network
(case 5)

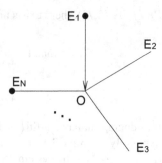

a C^1 solution $(A_N(t, x), V_N(t, x))$ on the domain $R_N(\bar{T})$, and, noting (5.64), by
Lemma 1.5 this solution satisfies the initial condition (5.5) for $i = N$.

Remark 5.7 In Theorem 5.5, both the number of the given nodal profiles and that of
the boundary controls still satisfy the relationship (4.60). In this situation, 2 nodal
profiles are given without constraint at the multiple node $x = 0$, while 2 boundary
controls are taken at $x = L_1$ and L_N, respectively. See Fig. 5.6.

Remark 5.8 Comparing cases 4 and 5 with cases 1 and 2, respectively, we find that
the boundary controls can transfer from one node to another one. In Theorems 5.1
and 5.2, there is always a boundary control at the multiple node $x = 0$, while, in
Theorems 5.4 and 5.5, this boundary control transfers from the multiple node $x = 0$
to a simple node $x = L_N$ (see Figs. 5.2 and 5.5, also see Figs. 5.3(2) and 5.6). This is
a special property for unsteady flows on a star-like network of open canals, caused
by the particular form of the interface conditions at the multiple node. Thus, for
unsteady flows on a star-like network of open canals, wherever the nodal profiles are
given, the boundary controls can be chosen only at simple nodes. So, in practical
problems, if the boundary control can not be chosen at the multiple node, namely,
if the total flux interface condition (5.7) is homogeneous, or, more generally, if the
total flux is given at the multiple node, the exact boundary controllability of nodal
profile can still be realized.

5.6   Exact Boundary Controllability of Nodal Profile
for Saint-Venant System on a Tree-Like Network
of Open Canals

In this section, we will improve and generalize the results given in Sects. 5.1–5.5 to
the case of tree-like network with general topology.

Consider a tree-like network composed of N horizontal and cylindrical canals:
$C_1, \cdots, C_N$. Without loss of generality, we suppose that one endpoint of canal C_1
is a simple node of the network, and take it as a starting node E. See Fig. 5.7.

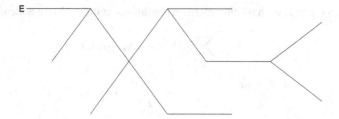

Fig. 5.7 Tree-like network of open canals

For the ith canal, let d_{i0} and d_{i1} be the x coordinates of its two endpoints, and let $L_i = d_{i1} - d_{i0}$ be its length. For the convenience of statement, in what follows we simply use d_{i0} and d_{i1} to express the nodes $x = d_{i0}$ and d_{i1}, respectively. We suppose that d_{i0} is closer to E than d_{i1} in the network and d_{10} is just E. The corresponding Saint-Venant system can be written as

$$\begin{cases} \dfrac{\partial A_i}{\partial t} + \dfrac{\partial(A_i V_i)}{\partial x} = 0, \\ \dfrac{\partial V_i}{\partial t} + \dfrac{\partial S_i}{\partial x} = 0, \end{cases} \quad t \geq 0, \quad d_{i0} \leq x \leq d_{i1} \quad (i = 1, \cdots, N), \quad (5.78)$$

where

$$S_i = \frac{1}{2} V_i^2 + g h_i(A_i) + g Y_{bi} \quad (i = 1, \cdots, N) \quad (5.79)$$

with

$$h_i'(A_i) > 0 \quad (i = 1, \cdots, N) \quad (5.80)$$

and Y_{bi} $(i = 1, \cdots, N)$ being constants.

The initial condition is given by

$$t = 0: \quad (A_i, V_i) = (A_{i0}(x), V_{i0}(x)), \quad d_{i0} \leq x \leq d_{i1} \quad (i = 1, \cdots, N). \quad (5.81)$$

Let $\mathcal{M}$ and $\mathcal{S}$ be two subsets of $\{1, \cdots, N\}$, such that $i \in \mathcal{M}$ if and only if d_{i1} is a multiple node, while, $i \in \mathcal{S}$ if and only if d_{i1} is a simple node. In other words, $\mathcal{S}$ is the index set of all the simple nodes except E, while $\mathcal{M}$ is the index set of all the multiple nodes.

At the simple node d_{10}, we have the flux boundary condition

$$x = d_{10}: \quad A_1 V_1 = q_0(t). \quad (5.82)$$

When $i \in \mathcal{S}$, we have the flux boundary condition at the simple node d_{i1}

$$x = d_{i1}: \quad A_i V_i = q_i(t), \quad (5.83)$$

while, when $i \in \mathcal{M}$, we have the interface conditions at the multiple node d_{i1}

$$\sum_{j \in \mathcal{J}_i} A_j V_j = A_i V_i + q_i(t), \qquad (5.84)$$

$x = d_{i1}:$

$$S_j = S_i, \quad \forall j \in \mathcal{J}_i, \qquad (5.85)$$

where $\mathcal{J}_i$ denotes the set of all the indices j such that node d_{j0} is just node d_{i1}, namely, $\mathcal{J}_i$ is the index set of all other canals which jointly possess the multiple node d_{i1} with the ith canal C_i.

Next, we will first generalize the previous results given in Sects. 5.1–5.3 for the star-like network to the tree-like network. Since only one function can be chosen as a boundary control at each multiple node (see (5.84)), the nodal profiles can not be given too arbitrarily. Thus, both nodal profiles and boundary controls should satisfy some principles.

Principle 5.1 (Principle of providing nodal profiles):   Nodal profiles should always be given by a pair of $(A_i, V_i) = (\bar{\bar{a}}_i(t), \bar{\bar{v}}_i(t))$. For any given canal C_i in the network, (A_i, V_i) as nodal profiles should be possibly given only at one node of this canal. For any given two canals with a joint node, for example, the ith and the kth canals, if (A_i, V_i) and (A_k, V_k) are simultaneously given as nodal profiles, then these two pairs of profiles should be given on the joint multiple node.

Correspondingly, we have

Principle 5.2 (Principle of choosing boundary controls): If a pair of nodal profiles is given at a simple node, then at another node (multiple one) of the corresponding canal, we choose $q_i(t)$ in the total flux interface condition (5.84) as a boundary control. If some pairs of nodal profiles are given at a multiple node, then (1) when these nodal profiles are not given on all the canals with this multiple node as their common node, we choose the total flux $q_i(t)$ in (5.84) at this multiple node as a boundary control, while, at other nodes (simple or multiple ones) of the canals corresponding to the given nodal profiles, we choose $q_i(t)$ in (5.82), (5.83) or (5.84) as boundary controls. (2) when these nodal profiles are given on all the canals with this multiple node as their common node, we choose $q_i(t)$ in (5.82), (5.83) or (5.84) at all other nodes of these canals as boundary controls.

Based on these principles, we can get the following theorem.

Theorem 5.6 *For any given tree-like network of open canals, there exists a suitably large controllability time T and $\bar{T} > T$, if some nodal profiles are given according to Principle 5.1 on the interval $[T, \bar{T}]$, then we can choose corresponding boundary controls according to Principle 5.2 on the interval $[0, \bar{T}]$, such that the exact boundary controllability of nodal profile can be realized on this network. Here, we still ask that the C^1 norms of the corresponding functions are suitably small and the conditions of piecewise C^1 compatibility are satisfied.*

Proof Similarly to the proof of Theorems 5.1, 5.2 and 5.3, on the tree-like network under consideration we need only to construct a piecewise C^1 solution to system (5.78) on the domain $R(\bar{T})$, such that it satisfies the initial condition (5.81), those boundary conditions and interface conditions (5.82)–(5.85) without boundary controls, and the nodal profiles given on the interval $[T, \bar{T}]$. For this purpose, we will use the induction on the number of the multiple nodes of the tree-like network.

For a network with only one multiple node, namely, for a star-like network, by Principles 5.1 and 5.2, the nodal profiles and the corresponding boundary controls should be given by the way described in Theorems 5.1, 5.2 and 5.3. So, Theorem 5.6 is true.

We now suppose that Theorem 5.6 is true for a tree-like network with M multiple nodes. We consider a network of open canals with $(M + 1)$ multiple nodes.

Situation 1: There exists a multiple node d_{i1} ($i \in \mathcal{M}$) at which some pairs of nodal profiles are given. Without loss of generality, we suppose that (A_i, V_i) is such a pair of nodal profiles given at d_{i1}. Cutting the network at the multiple node d_{i1}, we get several subnetworks, and the number of the multiple nodes in each subnetwork is fewer than M. See Fig. 5.8.

On the subnetwork with the ith canal (subnetwork A in Fig. 5.8), a pair of given nodal profiles $(A_i, V_i) = (\bar{\bar{a}}_i(t), \bar{\bar{v}}_i(t))$ is given at the simple node d_{i1} on the interval $[T, \bar{T}]$. It is easy to see that all the nodal profiles given on this subnetwork still satisfy Principle 5.1, and the corresponding boundary controls chosen according to Principle 5.2 on this subnetwork are exactly the same as the controls on the original network but restricted to the subnetwork. By induction, we can find a desired piecewise C^1 solution on this subnetwork. Thus, we can get $S_i(t)$, the value of S_i at $x = d_{i1}$, by the corresponding value of $(A_i(t, x), V_i(t, x))$.

First, we consider the case that the nodal profiles are not given at all the nodes d_{j0} for $j \in \mathcal{J}_i$. In this case, we consider a subnetwork containing the jth canal for $j \in \mathcal{J}_i$. At the simple node d_{j0} of this subnetwork, there may be no given nodal profiles (see subnetwork B) or a pair of nodal profiles (see subnetwork C). However, by Principle 5.1, the nodal profiles (A_j, V_j) should not be given at the node d_{j1} in both cases. Hence, with the boundary condition

$$S_j = S_i(t) \tag{5.86}$$

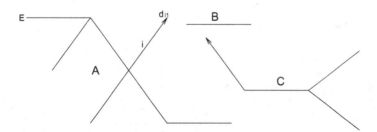

Fig. 5.8 Splitting the tree-like network (situation 1)

at $d_{j0} = d_{i1}$, by Remark 5.2 we can get a desired piecewise C^1 solution on this subnetwork by induction.

Considering all j ($j \in \mathcal{J}_i$) and putting all the piecewise C^1 solutions together, on the whole network we get a piecewise C^1 solution on the domain $R(\bar{T})$, which satisfies all the initial conditions and all the given nodal profiles. Since there is only a boundary control function in the total flux interface condition (5.84) at the multiple node d_{i1}, while, the "energy-type" interface conditions (5.85) are obviously satisfied, the solution satisfies all the boundary conditions and all the interface conditions without boundary controls on the whole network.

Next, we consider the case that the nodal profiles are given at all the nodes d_{j0} for $j \in \mathcal{J}_i$. In this case, any given subnetwork containing the jth canal for $j \in \mathcal{J}_i$ is similar to the subnetwork A, then we can easily get the same conclusion.

Situation 2: All the nodal profiles are given at simple nodes. Without loss of generality, we suppose that a pair of nodal profiles (A_1, V_1) are given at the simple node d_{10} on the interval $[T, \bar{T}]$. Thus, on the first canal, we can find a C^1 solution $(A_1(t, x), V_1(t, x))$ to system (5.78) for $i = 1$ with the initial condition (5.81) for $i = 1$, the boundary condition (5.82) and the nodal profiles given at the simple node d_{10}. Then we can get $S_1(t)$, the value of S_1 at the multiple node d_{11}, by the corresponding value of $(A_1(t, x), V_1(t, x))$.

Next, we cut the network at the multiple node d_{11} and get several subnetworks, and the number of the multiple nodes in each subnetwork is fewer than M. See Fig. 5.9.

We now consider the subnetwork containing the jth canal for $j \in \mathcal{J}_1$. At the simple node d_{j0} of the subnetwork, taking the boundary condition

$$S_j = S_1(t), \tag{5.87}$$

by Remark 5.2, we can similarly get a desired piecewise C^1 solution on this subnetwork by induction.

Considering all j ($j \in \mathcal{J}_1$), and putting all the piecewise C^1 solutions on these subnetworks together with $(A_1(t, x), V_1(t, x))$, on the whole network we get a piecewise C^1 solution on the domain $R(\bar{T})$, which satisfies all the initial conditions and all the given nodal profiles. Since there is only a boundary control function in the total flux interface condition (5.84) at the multiple node d_{11}, while, the "energy-type" interface conditions (5.85) are obviously satisfied, the solution satisfies all the

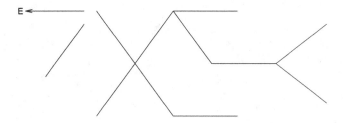

Fig. 5.9 Splitting the tree-like network (situation 2)

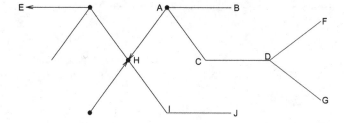

Fig. 5.10 Exact boundary controllability of nodal profile on a tree-like network

boundary conditions and all the interface conditions without boundary controls on the whole network. This finishes the proof.

Remark 5.9 By Principles 5.1 and 5.2, it is easy to see that (4.60) still holds for Theorem 5.6. See Fig. 5.10. There are 6 nodal profiles in total, 2 of them are given at the simple node E, but they should satisfy a corresponding boundary condition, while, the other 4 nodal profiles are given at the multiple node H, but they should satisfy a corresponding "energy-type" interface condition, so there are 2 constraints. Moreover, the number of the corresponding controls is equal to 4.

We now consider the possibility of transferring boundary control at a multiple node to that at a simple node.

By Theorems 5.4 and 5.5, we can give a more general principle for choosing boundary controls.

Principle 5.3 (Principle of choosing and transferring boundary controls): Based on choosing boundary controls according to Principle 5.2, suppose that we have chosen $q_i(t)$ as a boundary control at the multiple node d_{i1} ($i \in \mathcal{M}$), and there are neither nodal profiles nor boundary controls given at one of the nodes next to d_{i1}, for example, at d_{i0} (d_{k1} ($i \in \mathcal{J}_k$)) or at d_{j1} ($j \in \mathcal{J}_i$), then the original boundary control $q_i(t)$ can be replaced by $q_k(t)$ or $q_j(t)$, appearing in the total flux interface condition or in the flux boundary condition, at the corresponding node.

By Theorems 5.4 and 5.5, similarly to the proof of Theorem 5.6, we can get

Theorem 5.7 *For any given tree-like network of open canals, there exists a suitably large controllability time T and $\bar{T} > T$, if we give the nodal profiles according to Principle 5.1 on the interval $[T, \bar{T}]$, then we can choose the corresponding boundary controls according to Principles 5.2 and 5.3 on the interval $[0, \bar{T}]$, such that the exact boundary controllability of nodal profile can be realized on this tree-like network. Here, we still ask that the C^1 norms of the corresponding functions are suitably small and the conditions of piecewise C^1 compatibility are satisfied.*

By Theorem 5.7, the boundary controls chosen at node A according to Principle 5.2 in Fig. 5.10 can be replaced by the boundary controls at node B or C, see Figs. 5.11 and 5.12.

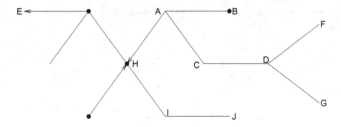

Fig. 5.11 Transfer of boundary controls (1)

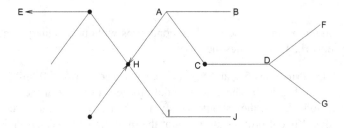

Fig. 5.12 Transfer of boundary controls (2)

Remark 5.10 By Principles 5.1, 5.2 and 5.3, it is easy to see that (4.60) still holds for Theorem 5.7.

Remark 5.11 Principle 5.3 can be successively applied. Suppose that a boundary control has been transferred and replaced by a new one according to Principle 5.3. If this new boundary control is still put at a multiple node and if there are neither nodal profiles nor boundary controls given at one of the nodes next to this multiple node, then the control can be further transferred to that node. In Fig. 5.12, the boundary control at node A has been transferred to node C according to Principle 5.3, then it can be further transferred to node D, F or G.

Moreover, several boundary controls on the network can be simultaneously transferred according to Principle 5.3. In Fig. 5.10, when we transfer the boundary control at node A as before, we can also transfer the boundary control at node H to node I or J. But the transfers of several boundary controls can not be interrupted by each other, in other words, all the canals involved in the transfer of a boundary control can not intersect with the canals involved in the transfer of another boundary control. See Fig. 5.13, the boundary controls at node A and K can not be simultaneously transferred to node L and J, respectively, since during the transfer of boundary controls, the canals intersect at joint node H.

Remark 5.12 If the number of the given nodal profiles is suitably small on a tree-like network of open canals, then we can get the exact boundary controllability of nodal profile with boundary controls only taken at some simple nodes by the transfer of boundary controls. In this situation, we can still get the controllability of nodal

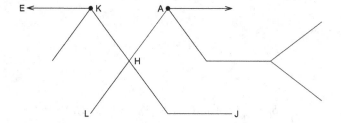

Fig. 5.13 Impossible transfer of boundary controls

profile, even though there are no boundary controls at multiple nodes, namely, the total flux interface condition (5.84) is homogeneous, or, more generally, the total flux is given at any fixed multiple node. Especially, for the situation that only one pair of nodal profiles is given on the tree-like network, if the nodal profiles are given at a simple node, then we can get the exact boundary controllability of nodal profile with one boundary control at any one of other simple nodes; while, if the nodal profiles are given at a multiple node, then we can get the exact boundary controllability of nodal profile with two boundary controls respectively at two simple nodes, and these two simple nodes should be on two subnetworks located respectively on two sides of the canal related to the given nodal profiles.

5.7   Exact Boundary Controllability of Nodal Profile for Saint-Venant System on a Tree-Like Network of Open Canals (Continued)

In this subsection, corresponding to the result given in Theorem 4.6, we discuss the case that certain nodal profiles are given at some internal nodes.

For this purpose, we first reconsider Theorem 4.6 from a different point of view,

The single open canal $[0, L]$ with an internal node $x = \theta L$ $(0 < \theta < 1)$ can be artificially regarded as a star-like network composed of two canals by taking this internal node as the multiple node. These two canals should have the same depth function of the water $h(A)$ and the same altitude of the bed of canal Y_b, moreover, the total flux function at this multiple node should be identically equal to zero.

The interface conditions (5.10)–(5.11) at this multiple node can be then written as

$$\sum_{i=1}^{2} A_i V_i = 0 \qquad (5.88)$$

and

$$S_1 = S_2, \qquad (5.89)$$

in which

$$S_i = \frac{1}{2}V_i^2 + gh(A_i) + gY_b \quad (i = 1, 2). \qquad (5.90)$$

It is easy to see that in the subcritical region, namely, when

$$|V| < \sqrt{gAh'(A)}, \qquad (5.91)$$

giving the nodal profiles $(A, V) = (\bar{\bar{a}}(t), \bar{\bar{v}}(t))$ at the internal node $x = \theta L$ for a single open canal is equivalent to give two pairs of nodal profiles $(\bar{\bar{a}}_i(t), \bar{\bar{v}}_i(t))$ $(i = 1, 2)$ satisfying (5.88)–(5.89) at the corresponding multiple node.

Thus, it is easy to check that Theorem 4.6 can be directly obtained by Theorem 5.3 for a star-like network with $N = 2$.

We now consider the exact boundary controllability of nodal profile for Saint-Venant system on a tree-like network of open canals with certain nodal profiles given at some internal nodes.

Without loss of generality we assume that a pair of nodal profiles are given only at one internal node on a canal. We can still regard this canal as a subnetwork composed of two canals with the internal node as the corresponding multiple node. These two canals should have the same $h(A)$ and Y_b, and the total flux function in the interface conditions satisfied at this artificial multiple node is identically equal to zero. In this way, the original tree-like network composed of N canals can be regarded as a tree-like network composed of $(N + 1)$ canals, in which two pairs of nodal profiles $(\bar{\bar{a}}_i(t), \bar{\bar{v}}_i(t))$ $(i = 1, 2)$, satisfying the interface conditions with the zero total flux function at the artificial multiple node, are given from both sides of the multiple node.

When the presence of these two pairs of nodal profiles satisfies Principle 5.1, by means of the corresponding boundary controls satisfying Principles 5.2 and 5.3, we can still get the exact boundary controllability of nodal profile for the original tree-like network with a pair of nodal profiles given at an internal node.

However, when the presence of these two pairs of nodal profiles does not satisfy Principle 5.1, for instance, there are already a pair of nodal profiles given at one of the two nodes of the corresponding canal, it is impossible to use this internal node to provide node profiles on it, and we have no exact boundary controllability of nodal profile on the original tree-like network with a pair of nodal profiles on this internal node.

Remark 5.13 (4.60) still holds in the case discussed in this subsection.

5.8 Remarks

For other physical models without zero eigenvalues, the exact boundary controllability of nodal profile on a tree-like network can be treated in a similar way, however, the physically meaningful interface conditions at the multiple nodes for each model should be carefully presented and considered. From this point of view, we need to precisely study each different model as we did for the Saint-Venant system in this chapter.

When the system under consideration possesses zero eigenvalues (for instance, 1.51 holds), the exact controllability of nodal profile is still an open problem up to now.

Chapter 6
Exact Boundary Controllability of Nodal Profile for 1-D Quasilinear Wave Equations

6.1 Introduction

The local exact boundary controllability for 1-D quasilinear wave equations has been obtained by means of a constructive method with modular structure (see [12,19]). In this Chapter, we will show that, based on the results given in Chap. 2, this constructive method can be elegantly modified to get the local exact boundary controllability of nodal profile for 1-D quasilinear wave equations (see Wang [21]).

In Sect. 6.2, we will give precise definitions and main results on the exact boundary controllability of nodal profile for the cases that the nodal profile is given on a boundary node and on an internal node, respectively, then the corresponding results of the local exact boundary controllability of nodal profile are proved in Sects. 6.3 and 6.4, respectively.

6.2 Definitions and Main Results

Consider the following 1-D quasilinear wave equation

$$\frac{\partial^2 u}{\partial t^2} - \frac{\partial}{\partial x}\left(K\left(u, \frac{\partial u}{\partial x}\right)\right) = F\left(u, \frac{\partial u}{\partial x}, \frac{\partial u}{\partial t}\right), \tag{6.1}$$

where t is the time variable, x is the spatial variable, u is unknown C^2 function of (t, x), $K = K(u, v)$ a given C^2 function with

$$K_v(u, v) > 0, \tag{6.2}$$

and, without loss of generality, we assume that

$$K(0, 0) = 0. \tag{6.3}$$

© The Author(s) 2016
T. Li et al., *Exact Boundary Controllability of Nodal Profile for Quasilinear Hyperbolic Systems*, SpringerBriefs in Mathematics,
DOI 10.1007/978-981-10-2842-7_6

Moreover, $F = F(u, v, w)$ is a given C^1 function, satisfying

$$F(0, 0, 0) = 0. \tag{6.4}$$

The initial condition is given as

$$t = 0: \quad u = \varphi(x), \ u_t = \psi(x), \quad 0 \le x \le L, \tag{6.5}$$

where L is the length of the spatial interval, $\varphi(x)$ and $\psi(x)$ are C^2 and C^1 functions, respectively, on the interval $0 \le x \le L$.

On one end $x = 0$, we prescribe any one of the following physically meaningful boundary conditions:

$x = 0: \ u = h(t)$	(Dirichlet type),	(6.6a)
$x = 0: \ u_x = h(t)$	(Neumann type),	(6.6b)
$x = 0: \ u_x - \alpha u = h(t)$	(Third type),	(6.6c)
$x = 0: \ u_x - \beta u_t = h(t)$	(Dissipative type),	(6.6d)

where α and β are given positive constants, $h \in C^2$ (for (6.6a)) or C^1 (for (6.6b)–(6.6d)) is a given function or a control function to be determined.

Similarly, on another end $x = L$, we prescribe any one of the following boundary conditions:

$x = L: \ u = \tilde{h}(t)$	(Dirichlet type),	(6.7a)
$x = L: \ u_x = \tilde{h}(t)$	(Neumann type),	(6.7b)
$x = L: \ u_x + \tilde{\alpha} u = \tilde{h}(t)$	(Third type),	(6.7c)
$x = L: \ u_x + \tilde{\beta} u_t = \tilde{h}(t)$	(Dissipative type),	(6.7d)

where $\tilde{\alpha}$ and $\tilde{\beta}$ are given positive constants, $\tilde{h} \in C^2$ (for (6.7a)) or C^1 (for (6.7b)–(6.7d)) is a given function or a control function to be determined.

For the convenience of statement, in what follows we denote that

$$\delta = \begin{cases} 2 & \text{in case (6.6a)} \\ 1 & \text{in cases (6.6b)–(6.6d)} \end{cases}, \quad \tilde{\delta} = \begin{cases} 2 & \text{in case (6.7a)} \\ 1 & \text{in cases (6.7b)–(6.7d)} \end{cases}. \tag{6.8}$$

Definition 6.1 (*Exact boundary controllability of nodal profile on a boundary node*) For any given initial data $(\varphi(x), \psi(x))$ and any given boundary function $\tilde{h}(t)$, satisfying the conditions of C^2 compatibility at the point $(t, x) = (0, L)$, for any given $C^2 \times C^1$ function $(\bar{\bar{u}}(t), \bar{v}(t))$, if there exist $T > 0$ and boundary control $h(t) \in C^\delta$, such that the C^2 solution $u = u(t, x)$ to the mixed initial-boundary value problem (6.1), (6.5), (6.6), and (6.7) fits exactly $(u, u_x) = (\bar{\bar{u}}(t), \bar{v}(t))$ on $x = L$ for $t \ge T$, then we have the exact boundary controllability of nodal profile on the boundary node $x = L$.

Remark 6.1 In Definition 6.1, when $t \geq T$, the given $C^2 \times C^1$ function $(\bar{\bar{u}}(t), \bar{\bar{v}}(t))$, being the value of (u, u_x) on $x = L$, should be consistent with the boundary condition (6.7), namely, it should be satisfied that

$$\bar{\bar{u}}(t) \equiv \tilde{h}(t), \qquad\qquad t \geq T \quad \text{(for (6.7a))}, \qquad (6.9a)$$

$$\bar{\bar{v}}(t) \equiv \tilde{h}(t), \qquad\qquad t \geq T \quad \text{(for (6.7b))}, \qquad (6.9b)$$

$$\bar{\bar{v}}(t) \equiv \tilde{h}(t) - \tilde{\alpha}\bar{\bar{u}}(t), \qquad t \geq T \quad \text{(for (6.7c))}, \qquad (6.9c)$$

$$\bar{\bar{v}}(t) \equiv \tilde{h}(t) - \tilde{\beta}\bar{\bar{u}}'(t), \qquad t \geq T \quad \text{(for (6.7d))}. \qquad (6.9d)$$

We will prove the following theorem in Sect. 6.3.

Theorem 6.1 *Let*

$$T > \frac{L}{\sqrt{K_v(0,0)}} \qquad (6.10)$$

and let $\bar{T}$ be an arbitrarily given number such that

$$\bar{T} > T. \qquad (6.11)$$

For any given initial data $(\varphi(x), \psi(x))$ and boundary function $\tilde{h}(t)$ with small norms $\|(\varphi, \psi)\|_{C^2[0,L] \times C^1[0,L]}$ and $\|\tilde{h}\|_{C^3[0,\bar{T}]}$, satisfying the conditions of C^2 compatibility at the point $(t, x) = (0, L)$, for any given values $\bar{\bar{u}}(t)$ and $\bar{\bar{v}}(t)$ satisfying the requirement in Remark 6.1 and having small norm $\|(\bar{\bar{u}}(t), \bar{\bar{v}}(t))\|_{C^2[T,\bar{T}] \times C^1[T,\bar{T}]}$, there exists a boundary control $h(t)$ with small norm $\|h(t)\|_{C^3[0,\bar{T}]}$, such that the mixed initial-boundary value problem (6.1), (6.5), (6.6), and (6.7) admits a unique C^2 solution $u = u(t, x)$ with small C^2 norm on the domain $R(\bar{T}) = \{(t, x)|0 \leq t \leq \bar{T}, 0 \leq x \leq L\}$, which fits exactly the given profiles $(u, u_x) = (\bar{\bar{u}}(t), \bar{\bar{v}}(t))$ on the boundary node $x = L$ for $T \leq t \leq \bar{T}$.

Remark 6.2 The exact boundary controllability of nodal profile on the boundary node $x = 0$ can be similarly defined and realized.

We next consider the exact boundary controllability of nodal profile on an internal node.

Definition 6.2 (*Exact boundary controllability of nodal profile on an internal node*) For any given initial data $(\varphi(x), \psi(x))$ and any given internal node $x = \theta L$ ($0 < \theta < 1$), for any given $C^2 \times C^1$ functions $(\bar{\bar{u}}(t), \bar{\bar{v}}(t))$, if there exist $T_\theta > 0$ and boundary controls $h(t)$ and $\tilde{h}(t)$, such that the C^2 solution $u = u(t, x)$ to the mixed initial-boundary value problem (6.1), (6.5), (6.6), and (6.7) fits exactly $(u, u_x) = (\bar{\bar{u}}(t), \bar{\bar{v}}(t))$ on $x = \theta L$ for $t \geq T_\theta$, then we have the exact boundary controllability of nodal profile on the internal node $x = \theta L$.

We will prove the following theorem in Sect. 6.4.

Theorem 6.2 *Let*

$$T_\theta > \frac{L}{\sqrt{K_v(0,0)}} \max(1-\theta, \theta), \tag{6.12}$$

in which $0 < \theta < 1$, and let $\bar{T}$ be an arbitrarily given number such that

$$\bar{T} > T_\theta. \tag{6.13}$$

For any given initial data $(\varphi(x), \psi(x))$ with small norm $\|(\varphi, \psi)\|_{C^2[0,L] \times C^1[0,L]}$, for any given values $(\bar{\bar{u}}(t), \bar{\bar{v}}(t))$ on $x = \theta L$ for $T_\theta \le t \le \bar{T}$ with small norm $\|(\bar{\bar{u}}(t), \bar{\bar{v}}(t))\|_{C^2[T_\theta, \bar{T}] \times C^1[T_\theta, \bar{T}]}$, there exist boundary controls $h(t)$ and $\tilde{h}(t)$ with small norms $\|h(t)\|_{C^5[0, \bar{T}]}$ and $\|\tilde{h}(t)\|_{C^5[0, \bar{T}]}$, respectively, such that the mixed initial-boundary value problem (6.1), (6.5), (6.6), and (6.7) admits a unique C^2 solution $u = u(t, x)$ with small C^2 norm on the domain $R(\bar{T}) = \{(t, x)|0 \le t \le \bar{T}, 0 \le x \le L\}$, which fits exactly the given profiles $(u, u_x) = (\bar{\bar{u}}(t), \bar{\bar{v}}(t))$ on the internal node $x = \theta L$ for $T_\theta \le t \le \bar{T}$.

Remark 6.3 When $\theta = 1/2$, namely, the internal node $x = \theta L$ is the middle point of the interval $[0, L]$, the right-hand side of (6.11) reaches the minimum, then we get the optimal controllability time, and (6.12) becomes

$$T_{\frac{1}{2}} > \frac{L}{2\sqrt{K_v(0,0)}}. \tag{6.14}$$

6.3   Proof of Theorem 6.1

To prove Theorem 6.1, we suitably modify the constructive method suggested in [12, 14, 16–18], and divide the proof in several steps.

(1) Noting (6.10), there exists an $\varepsilon_0 > 0$ so small that

$$T_1 < T, \tag{6.15}$$

where

$$T_1 = L \max_{|u|+|v| \le \varepsilon_0} \frac{1}{\sqrt{K_v(u,v)}}. \tag{6.16}$$

On the domain $R(T_1) = \{(t, x)|0 \le t \le T_1, 0 \le x \le L\}$, we consider a forward mixed initial-boundary value problem for Eq. (6.1) with the initial condition (6.5), the boundary condition (6.7) on $x = L$ and the following artificial boundary condition:

$$x = 0: \quad u = f(t), \tag{6.17}$$

where $f(t)$ is an arbitrarily given C^2 function with small $C^2[0, T_1]$ norm, satisfying the conditions of C^2 compatibility at the point $(t, x) = (0, 0)$.

By Lemma 2.1, this forward mixed initial-boundary value problem admits a unique C^2 solution $u = u_f(t, x)$ with small C^2 norm on the domain $R(T_1)$. In particular, we have

$$|u_f(t, x)| + \left|\frac{\partial u_f}{\partial x}(t, x)\right| \leq \varepsilon_0, \quad \forall(t, x) \in R(T_1). \tag{6.18}$$

Thus, we can uniquely determine the value of (u_f, u_{fx}) on $x = L$ for $0 \leq t \leq T_1$:

$$x = L: \quad (u_f, u_{fx}) = (\bar{u}(t), \bar{v}(t)), \quad 0 \leq t \leq T_1, \tag{6.19}$$

and its $C^2[0, T_1] \times C^1[0, T_1]$ norm is small.

By (6.15), there exists $(u(t), v(t)) \in C^2[0, \bar{T}] \times C^1[0, \bar{T}]$ with small $C^2 \times C^1$ norm, such that

$$(u(t), v(t)) = \begin{cases} (\bar{u}(t), \bar{v}(t)), & 0 \leq t \leq T_1 \\ (\bar{\bar{u}}(t), \bar{\bar{v}}(t)), & T \leq t \leq \bar{T} \end{cases} \tag{6.20}$$

and $(u(t), v(t))$ satisfies the boundary condition (6.7) on the whole interval $0 \leq t \leq \bar{T}$.

(2) We exchange the role of t and x, and consider a leftward mixed initial-boundary value problem on the domain $R(\bar{T})$ for Eq. (6.1) with the initial condition:

$$x = L: \quad (u, u_x) = (u(t), v(t)), \quad 0 \leq t \leq \bar{T}, \tag{6.21}$$

the boundary condition reduced from the original initial condition (6.5)

$$t = 0: \quad u = \varphi(x), \quad 0 \leq x \leq L \tag{6.22}$$

and the following artificial boundary condition:

$$t = \bar{T}: \quad u = \Phi(x), \quad 0 \leq x \leq L, \tag{6.23}$$

where $\Phi(x)$ is an arbitrarily given C^2 function with small $C^2[0, L]$ norm, satisfying the conditions of C^2 compatibility at the point $(t, x) = (\bar{T}, L)$.

By Lemma 2.1, this leftward mixed problem admits a unique C^2 solution $u = u(t, x)$ with small C^2 norm on $R(\bar{T})$. In particular, we have

$$|u(t, x)| + |u_x(t, x)| \leq \varepsilon_0, \quad \forall(t, x) \in R(\bar{T}). \tag{6.24}$$

(3) This C^2 solution $u = u(t, x)$ satisfies Eq. (6.1), the first initial condition (6.22) and the boundary condition (6.7). Next, we prove that it also satisfies the second initial condition

$$t = 0: \quad u_t = \psi(x). \tag{6.25}$$

In fact, consider the following one-sided mixed initial-boundary value problem for Eq. (6.1) with the initial condition

$$x = L: \quad (u, u_x) = (\bar{u}(t), \bar{v}(t)), \quad 0 \le t \le T_1 \tag{6.26}$$

and the boundary condition (6.22). Both $u = u(t, x)$ and $u = u_f(t, x)$ are C^2 solutions to this one-sided mixed problem. By the choice of T_1 in (6.16), noting (6.18) and (6.24), by Lemma 2.3, it is easy to see that the interval $0 \le x \le L$ on the initial axis $t = 0$ is included in the maximum determinate domain of this one-sided mixed problem, thus, by the uniqueness of C^2 solution to the one-sided mixed problem, $u(t, x) \equiv u_f(t, x)$ on this interval $\{t = 0, 0 \le x \le L\}$, then $u = u(t, x)$ satisfies (6.25).

Finally, substituting $u = u(t, x)$ into the boundary condition (6.6) on $x = 0$, we get the desired boundary control function $h(t)$ for $0 \le t \le \bar{T}$.

6.4 Proof of Theorem 6.2

Similarly to the proof of Theorem 6.1, we divide the proof into several steps.

(1) Noting (6.12), there exists an $\varepsilon_0 > 0$ so small that

$$T_{1\theta} < T_\theta, \tag{6.27}$$

where

$$T_{1\theta} = L \max_{|u|+|v| \le \varepsilon_0} \left(\frac{1-\theta}{\sqrt{K_v(u, v)}}, \frac{\theta}{\sqrt{K_v(u, v)}} \right). \tag{6.28}$$

On the domain $R(T_{1\theta}) = \{(t, x)|0 \le t \le T_{1\theta}, 0 \le x \le L\}$, we consider a forward mixed initial-boundary value problem for Eq. (6.1) with the initial condition (6.5) and the following artificial boundary conditions

$$x = 0: \quad u = f(t), \tag{6.29}$$
$$x = L: \quad u = g(t), \tag{6.30}$$

where $f(t)$ and $g(t)$ are arbitrarily given C^2 functions with small $C^2[0, T_{1\theta}]$ norms, satisfying the conditions of C^2 compatibility at the points $(t, x) = (0, 0)$ and $(0, L)$, respectively.

By Lemma 2.1, this forward mixed problem admits a unique C^2 solution $u = u_f(t, x)$ with small C^2 norm on the domain $R(T_{1\theta})$. In particular, we have

$$|u_f(t, x)| + \left|\frac{\partial u_f}{\partial x}(t, x)\right| \le \varepsilon_0, \quad \forall (t, x) \in R(T_{1\theta}). \tag{6.31}$$

Thus, we can uniquely determine the value of (u_f, u_{fx}) on $x = \theta L$ for $0 \le t \le T_{1\theta}$,

$$x = \theta L : \quad (u_f, u_{fx}) = (\bar{u}(t), \bar{v}(t)), \quad 0 \le t \le T_{1\theta}, \tag{6.32}$$

and its $C^2[0, T_{1\theta}] \times C^1[0, T_{1\theta}]$ norm is small.

By (6.27), there exists $(u(t), v(t)) \in C^2[0, \bar{T}] \times C^1[0, \bar{T}]$ with small $C^2 \times C^1$ norm, such that

$$(u(t), v(t)) = \begin{cases} (\bar{u}(t), \bar{v}(t)), & 0 \le t \le T_{1\theta}, \\ (\bar{\bar{u}}(t), \bar{\bar{v}}(t)), & T_\theta \le t \le \bar{T}. \end{cases} \tag{6.33}$$

(2) We exchange the role of t and x, and consider a leftward (resp. rightward) mixed initial-boundary value problem on $R_l(\bar{T}) = \{(t, x)|0 \le t \le \bar{T}, 0 \le x \le \theta L\}$ (resp. $R_r(\bar{T}) = \{(t, x)|0 \le t \le \bar{T}, \theta L \le x \le L\}$) for Eq. (6.1) with the initial condition

$$x = \theta L : \quad (u, u_x) = (u(t), v(t)), \quad 0 \le t \le \bar{T}, \tag{6.34}$$

the boundary condition reduced from the original initial condition (6.5)

$$t = 0 : \quad u = \varphi(x), \quad 0 \le x \le \theta L \tag{6.35}$$
$$(\text{resp. } t = 0 : \quad u = \varphi(x), \quad \theta L \le x \le L)$$

and the following artificial boundary condition

$$t = \bar{T} : \quad u = \Phi(x), \quad 0 \le x \le \theta L \tag{6.36}$$
$$(\text{resp. } t = \bar{T} : \quad u = \Phi(x), \quad \theta L \le x \le L),$$

where $\Phi(x)$ is an arbitrarily given C^2 function with small $C^2[0, \theta L]$ norm (resp. $C^2[\theta L, L]$ norm), satisfying the conditions of C^2 compatibility at the point $(t, x) = (\bar{T}, \theta L)$.

By Lemma 2.1, this leftward (resp. rightward) mixed problem admits a unique C^2 solution $u = u_l(t, x)$ (resp. $u = u_r(t, x)$) with small C^2 norm on $R_l(\bar{T})$ (resp. $R_r(\bar{T})$). In particular, we have

$$|u_l(t, x)| + |u_{lx}(t, x)| \le \varepsilon_0, \quad \forall (t, x) \in R_l(\bar{T}) \tag{6.37}$$
$$(\text{resp. } |u_r(t, x)| + |u_{rx}(t, x)| \le \varepsilon_0, \quad \forall (t, x) \in R_r(\bar{T})).$$

(3) Let

$$u(t, x) = \begin{cases} u_l(t, x), & (t, x) \in R_l(\bar{T}), \\ u_r(t, x), & (t, x) \in R_r(\bar{T}). \end{cases} \tag{6.38}$$

Then $u = u(t, x)$ is a C^2 solution to Eq. (6.1) on the whole domain $R(\bar{T})$, satisfying the first initial condition (6.35). Next we prove that it also satisfies the second initial condition

$$t = 0: \quad u_t = \psi(x), \quad 0 \le x \le L. \tag{6.39}$$

For this purpose, we consider the following leftward (resp. rightward) one-sided mixed initial-boundary value problem for Eq. (6.1) with the initial condition

$$x = \theta L: \quad (u, u_x) = (\bar{u}(t), \bar{v}(t)), \quad 0 \le t \le T_{1\theta} \tag{6.40}$$

and the boundary condition (6.35). Both $u = u(t, x)$ and $u = u_f(t, x)$ are C^2 solutions to these one-sided mixed problems for $0 \le t \le T_{1\theta}$. By the choice (6.28) of $T_{1\theta}$, noting (6.31) and (6.37), by Lemma 2.3 it is easy to see that the interval $0 \le x \le \theta L$ (resp. $\theta L \le x \le L$) on the initial axis $t = 0$ is included in the corresponding maximum determinate domain of the corresponding one-sided mixed problem. Thus, by the uniqueness of C^2 solution to the one-sided mixed problem, $u(t, x) \equiv u_f(t, x)$ on this interval $\{t = 0, 0 \le x \le L\}$, then $u = u(t, x)$ satisfies (6.39).

Finally, substituting $u = u(t, x)$ into the boundary conditions (6.6) and (6.7), we get the desired boundary control functions $h(t)$ and $\bar{h}(t)$ for $0 \le t \le \bar{T}$.

6.5   Remarks

For general 1-D quasilinear hyperbolic equations (systems) of second order, the exact boundary controllability of nodal profile can be obtained in a similar way, provided that there are no zero eigenvalues.

The case with zero eigenvalues, which should be considered by means of not only boundary controls but also internal controls, is still open up to now.

Chapter 7
Exact Boundary Controllability of Nodal Profile for 1-D Quasilinear Wave Equations on a Planar Tree-Like Network of Strings

7.1 Introduction

In this Chapter, we will generalize the exact boundary controllability of nodal profile for 1-D quasilinear wave equations in a single string, discussed in Chap. 6, to that on a planar tree-like network of strings with general topology (see Wang and Gu [22]. For the corresponding result on the exact boundary controllability, cf. Gu and Li [6]).

In Sects. 7.2 and 7.3, we will first discuss the exact boundary controllability of nodal profile for 1-D quasilinear wave equations on a planar star-like network of strings with nodal profiles given at a simple node and at the multiple node, respectively. Then, we generalize these results to a tree-like network of strings in Sects. 7.4 and 7.5, in which we will present the principles of providing nodal profiles, choosing and transferring boundary controls in a tree-like network of strings with general topology. Based on these principles, by an induction, we get the exact boundary controllability of nodal profile for 1-D quasilinear wave equations on a planar tree-like network of strings with general topology, moreover, the relationship between the number of given nodal profiles and the number of required boundary controls is also given.

7.2 Exact Boundary Controllability of Nodal Profile for 1-D Quasilinear Wave Equations on a Star-Like Network of Strings

In this section and in the next section, we will consider the exact boundary controllability of nodal profile on a planar star-like network of strings.

As in Chap. 5, a star-like network is a connected network with only one multiple node. Suppose that the star-like network is composed of N strings with the joint

© The Author(s) 2016
T. Li et al., *Exact Boundary Controllability of Nodal Profile for Quasilinear Hyperbolic Systems*, SpringerBriefs in Mathematics, DOI 10.1007/978-981-10-2842-7_7

node O, and E_i is another node of the ith string with length L_i ($i = 1, \ldots, N$). Let the coordinate of O be $x = 0$. The ith string can be parameterized as $x \in [0, L_i]$ for $i = 1, \ldots, N$.

The 1-D quasilinear wave equations on this star-like network can be described as

$$\frac{\partial^2 u_i}{\partial t^2} - \frac{\partial}{\partial x}\left(K_i\left(u_i, \frac{\partial u_i}{\partial x}\right)\right) = F_i\left(u_i, \frac{\partial u_i}{\partial x}, \frac{\partial u_i}{\partial t}\right) \quad (i = 1, \ldots, N), \qquad (7.1)$$

where t is the time variable, x is the spatial variable and for each $i = 1, \ldots, N$, u_i is the unknown C^2 function of (t, x), $K_i = K_i(u, v)$ is a given C^2 function satisfying

$$K_{iv}(u, v) > 0, \qquad (7.2)$$

and, without loss of generality, we assume that

$$K_i(0, 0) = 0; \qquad (7.3)$$

moreover, $F_i = F_i(u, v, w)$ is a given C^1 function with

$$F_i(0, 0, 0) = 0. \qquad (7.4)$$

The initial condition is given as

$$t = 0: \quad u_i = \varphi_i(x), \quad u_{it} = \psi_i(x), \quad 0 \le x \le L_i \quad (i = 1, \ldots, N). \qquad (7.5)$$

For $i = 1, \ldots, N$, at each simple node E_i, we give one of the following physically meaningful boundary conditions:

$$x = L_i: \ u_i = h_i(t) \qquad \text{(Dirichlet type)}, \qquad (7.6a)$$
$$x = L_i: \ u_{ix} = h_i(t) \qquad \text{(Neumann type)}, \qquad (7.6b)$$
$$x = L_i: \ u_{ix} + \alpha_i u_i = h_i(t) \qquad \text{(Third type)}, \qquad (7.6c)$$
$$x = L_i: \ u_{ix} + \beta_i u_{it} = h_i(t) \qquad \text{(Dissipative type)}, \qquad (7.6d)$$

where α_i and β_i are given positive constants, $h_i(t) \in C^2$ (for (7.6a)) or C^1 (for (7.6b)–(7.6d)) is a given boundary function or a control function to be determined.

At the multiple node O, we prescribe the following interface conditions:

$$\sum_{i=1}^{N} K_i(u_i, u_{ix}) = h_0(t) \qquad (7.7)$$

and

$$u_i = u_1 \quad (i = 2, \ldots, N), \qquad (7.8)$$

where $h_0(t) \in C^1$. Equation (7.7) means that the total stress at O is equal to $h_0(t)$, and (7.8) shows the continuity of displacements at O.

For the convenience of statement, in what follows we denote that

$$\delta = \begin{cases} 2 & \text{for (7.6a),} \\ 1 & \text{for (7.6b)} - (7.6d). \end{cases} \tag{7.9}$$

Theorem 7.1 *Let*

$$T > \frac{L_1}{\sqrt{K_{1v}(0,0)}} \tag{7.10}$$

and $\bar{T}$ be an arbitrarily given number such that

$$\bar{T} > T. \tag{7.11}$$

For any given initial data $(\varphi_i(x), \psi_i(x))$ and for any given boundary functions $h_i(t)$ with small norms $\|(\varphi_i(\cdot), \psi_i(\cdot))\|_{C^2[0,L] \times C^1[0,L]}$ and $\|h_i(\cdot)\|_{C^\delta[0,\bar{T}]}$ for $i = 1, \dots, N$, such that the conditions of C^2 compatibility are satisfied at the points $(t, x) = (0, L_i)$ $(i = 1, \dots, N)$, respectively, for any given values $(\bar{\bar{u}}_1(t), \bar{\bar{v}}_1(t))$, being the value of (u_1, u_{1x}) at $x = L_1$, on the interval $[T, \bar{T}]$ with small norm $\|(\bar{\bar{u}}_1(\cdot), \bar{\bar{v}}_1(\cdot))\|_{C^2[T,\bar{T}] \times C^1[T,\bar{T}]}$, and satisfying the boundary condition (7.6) (in which $i = 1$) at the simple node E_1, there exists a boundary control $h_0(t)$ at the joint node O with small norm $\|h_0(\cdot)\|_{C^1[0,\bar{T}]}$, such that the mixed initial-boundary value problem (7.1) and (7.5)–(7.8) admits a unique piecewise C^2 solution $u_i = u_i(t, x)$ $(i = 1, \dots, N)$ with small piecewise C^2 norm on the domain $R(\bar{T}) = \bigcup\limits_{i=1}^{N} R_i(\bar{T}) = \bigcup\limits_{i=1}^{N} \{(t, x) | 0 \le t \le \bar{T}, 0 \le x \le L_i\}$, which fits exactly the given values $(u_1, u_{1x}) = (\bar{\bar{u}}_1(t), \bar{\bar{v}}_1(t))$ at the simple node $x = L_1$ for $T \le t \le \bar{T}$.

The proof of Theorem 7.1 is similar to that of Theorem 5.1 in Chap. 5. Theorem 7.1 can be illustrated by Fig. 7.1, here and hereafter, the symbol "→" means providing a node profile or a pair of nodal profiles with constraint at this node, while the symbol "•" means choosing a control at this node.

In Theorem 7.1 we choose the total stress function $h_0(t)$ at the joint node O as the control function (**case 1**). The following theorem shows that we can also choose the boundary function at another simple node as the control function (**case 2**).

Theorem 7.2 *Let*

$$T > \frac{L_1}{\sqrt{K_{1v}(0,0)}} + \frac{L_N}{\sqrt{K_{Nv}(0,0)}} \tag{7.12}$$

Fig. 7.1 Exact boundary controllability of nodal profile on a star-like network with nodal profiles given at a simple node (case 1)

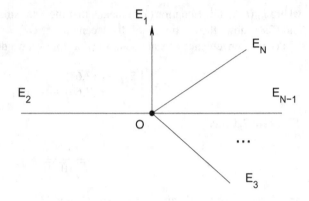

and $\bar{T}$ be an arbitrarily given number such that

$$\bar{T} > T. \tag{7.13}$$

For any given initial data $(\varphi_i(x), \psi_i(x))$ $(i = 1, \ldots, N)$ and for any given functions $h_0(t)$ and $h_j(t)$ $(j = 1, \ldots, N - 1)$ with small norms $\|(\varphi_i(\cdot), \psi_i(\cdot))\|_{C^2[0,L] \times C^1[0,L]}$ $(i = 1, \ldots, N)$, $\|h_0(\cdot)\|_{C^1[0,\bar{T}]}$ and $\|h_j(\cdot)\|_{C^\delta[0,\bar{T}]}$ $(j = 1, \ldots, N - 1)$, such that the conditions of piecewise C^2 compatibility are satisfied at the points $(t, x) = (0, 0)$ and $(0, L_j)$ $(j = 1, \ldots, N - 1)$, respectively, for any given values $(\bar{\bar{u}}_1(t), \bar{\bar{v}}_1(t))$, being the value of (u_1, u_{1x}) at $x = L_1$, on the interval $[T, \bar{T}]$ with small norm $\|(\bar{\bar{u}}_1(\cdot), \bar{\bar{v}}_1(\cdot))\|_{C^2[T,\bar{T}] \times C^1[T,\bar{T}]}$, and satisfying the boundary condition (7.6) (in which $i = 1$) at the simple node E_1, there exists a boundary control $h_N(t)$ with small norm $\|h_N(\cdot)\|_{C^\delta[0,\bar{T}]}$, such that the mixed initial-boundary value problem (7.1) and (7.5)–(7.8) admits a unique piecewise C^2 solution $u_i = u_i(t, x)$ $(i = 1, \ldots, N)$ with small piecewise C^2 norm on the domain $R(\bar{T}) = \bigcup\limits_{i=1}^{N} R_i(\bar{T})$, which fits exactly the given values $(u_1, u_{1x}) = (\bar{\bar{u}}_1(t), \bar{\bar{v}}_1(t))$ at the simple node $x = L_1$ for $T \le t \le \bar{T}$ (see Fig. 7.2).

Proof We prove it by a constructive method.

(1) Noting (7.12), there exists an $\varepsilon_0 > 0$ so small that

$$T_1 \stackrel{\text{def.}}{=} \max_{|u_1|+|v_1| \le \varepsilon_0} \frac{L_1}{\sqrt{K_{1v}(u_1, v_1)}} + \max_{|u_N|+|v_N| \le \varepsilon_0} \frac{L_N}{\sqrt{K_{Nv}(u_N, v_N)}} < T. \tag{7.14}$$

Let

$$T_2 = \max_{|u_N|+|v_N| \le \varepsilon_0} \frac{L_N}{\sqrt{K_{Nv}(u_N, v_N)}}. \tag{7.15}$$

Fig. 7.2 Exact boundary controllability of nodal profile on a star-like network with nodal profiles given at a simple node (case 2)

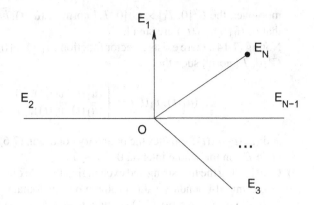

For this star-like network of strings, on the domain $R(T_1) = \bigcup\limits_{i=1}^{N} R_i(T_1) = \bigcup\limits_{i=1}^{N} \{(t, x)|0 \le t \le T_1, 0 \le x \le L_i\}$, we first consider a forward mixed initial-boundary value problem for system (7.1) with the initial condition (7.5), the interface conditions (7.7)–(7.8) at the multiple node O, the boundary condition (7.6) at the simple node E_j $(j = 1, \ldots, N - 1)$ and the following artificial boundary condition at the simple node E_N

$$x = L_N : \quad u_N = f(t), \tag{7.16}$$

where $f(t)$ is a C^2 function with small $C^2[0, T_1]$ norm and satisfies the conditions of C^2 compatibility at the point $(t, x) = (0, L_N)$.

By Lemma 3.2, this forward mixed problem admits a unique piecewise C^2 solution $u_i = u_{if}(t, x)$ $(i = 1, \ldots, N)$ on the domain $R(T_1)$ with a small piecewise C^2 norm. In particular, we have

$$\left| u_{if}(t, x) \right| + \left| \frac{\partial u_{if}}{\partial x}(t, x) \right| \le \varepsilon_0, \quad \forall (t, x) \in R_i(T_1) \quad (i = 1, \ldots, N). \tag{7.17}$$

Thus, we can uniquely determine the corresponding value of $\left(u_{if}, \frac{\partial u_{if}}{\partial x} \right)$ $(i = 1, \ldots, N)$ at $x = 0$ and that of $\left(u_{1f}, \frac{\partial u_{1f}}{\partial x} \right)$ at $x = L_1$ on the interval $[0, T_1]$ as follows:

$$x = 0 : \quad \left(u_{if}, \frac{\partial u_{if}}{\partial x} \right) = (\bar{a}_i(t), \bar{b}_i(t)), \quad 0 \le t \le T_1, \quad i = 1, \ldots, N, \tag{7.18}$$

$$x = L_1 : \quad \left(u_{1f}, \frac{\partial u_{1f}}{\partial x} \right) = (\bar{u}_1(t), \bar{v}_1(t)), \quad 0 \le t \le T_1, \tag{7.19}$$

moreover, the $C^2[0, T_1] \times C^1[0, T_1]$ norm of $(\bar{a}_i(t), \bar{b}_i(t))$ $(i = 1, \ldots, N)$ and that of $(\bar{u}_1(t), \bar{v}_1(t))$ are small.

Noting (7.14), there exists a vector function $(u_1(t), v_1(t))$ with small $C^2[0, \bar{T}] \times C^1[0, \bar{T}]$ norm, such that

$$(u_1(t), v_1(t)) = \begin{cases} (\bar{u}_1(t), \bar{v}_1(t)), & 0 \le t \le T_1, \\ (\hat{\bar{u}}_1(t), \hat{\bar{v}}_1(t)), & T \le t \le \bar{T}, \end{cases} \tag{7.20}$$

and $(u_1(t), v_1(t))$ satisfies the boundary condition (7.6) (for $i = 1$) at the simple node E_1 on the whole interval $0 \le t \le \bar{T}$.

(2) Considering the first string and exchanging the role of t and x, we solve a leftward mixed initial-boundary value problem on the domain $R_1(\bar{T}) = \{(t, x)|0 \le t \le \bar{T}, 0 \le x \le L_1\}$ for Eq. (7.1) (in which we take $i = 1$) together with the initial condition

$$x = L_1: \quad (u_1, u_{1x}) = (u_1(t), v_1(t)), \quad 0 \le t \le \bar{T}, \tag{7.21}$$

the boundary condition reduced from the original initial condition (7.5)

$$t = 0: \quad u_1 = \varphi_1(x), \quad 0 \le x \le L_1 \tag{7.22}$$

and an artificial boundary condition

$$t = \bar{T}: \quad u_1 = \tilde{\varphi}_1(x), \quad 0 \le x \le L_1, \tag{7.23}$$

where $\tilde{\varphi}_1(x)$ is a C^2 function with small $C^2[0, L_1]$ norm, which satisfies the conditions of C^2 compatibility at the point $(t, x) = (\bar{T}, L_1)$.

By Lemma 2.1, this leftward mixed problem admits a unique C^2 solution $u_1 = u_1(t, x)$ with small C^2 norm on the domain $R_1(\bar{T})$. In particular, we have

$$\left| u_1(t, x) \right| + \left| \frac{\partial u_1}{\partial x}(t, x) \right| \le \varepsilon_0, \quad \forall (t, x) \in R_1(\bar{T}). \tag{7.24}$$

Thus, we can uniquely determine the value of (u_1, u_{1x}) at $x = 0$ as

$$x = 0: \quad (u_1, u_{1x}) = (a_1(t), b_1(t)), \quad 0 \le t \le \bar{T}. \tag{7.25}$$

Both the C^2 solutions $u_1 = u_1(t, x)$ and $u_1 = u_{1f}(t, x)$ satisfy simultaneously the one-sided mixed initial-boundary value problem for Eq. (7.1) (in which we take $i = 1$), the initial condition

$$x = L_1: \quad (u_1, u_{1x}) = (\bar{u}_1(t), \bar{v}_1(t)), \quad 0 \le t \le T_1 \tag{7.26}$$

and the boundary condition (7.22). Noting (7.14)–(7.15), (7.17) and (7.24), by Lemma 2.3, the maximum determinate domain of the solution to this one-sided mixed problem includes the domain $\{(t, x)|0 \leq t \leq T_2 + \frac{T_1 - T_2}{L_1}x, 0 \leq x \leq L_1\}$, which means $u_1(t, x) \equiv u_{1f}(t, x)$ on this domain. Thus, on the interval $\{t = 0, 0 \leq x \leq L_1\}$, $u_1 = u_1(t, x)$ satisfies the initial condition (7.5) (for $i = 1$); moreover, on the interval $\{x = 0, 0 \leq t \leq T_2\}$, we have

$$(a_1(t), b_1(t)) \equiv (\bar{a}_1(t), \bar{b}_1(t)), \quad 0 \leq t \leq T_2. \tag{7.27}$$

(3) Based on the solution $u_1 = u_1(t, x)$ on the domain $R_1(\bar{T})$, obtained in step (2), by the interface condition (7.7) at the multiple node $x = 0$, we obtain

$$x = 0: \quad u_2 = \ldots = u_N = a_1(t), \quad 0 \leq t \leq \bar{T}, \tag{7.28}$$

in particular, we have

$$x = 0: \quad u_2 = \ldots = u_{N-1} = a_1(t), \quad 0 \leq t \leq \bar{T}. \tag{7.29}$$

For each $i = 2, \ldots, N - 1$, we consider a forward mixed initial-boundary value problem on the domain $R_i(\bar{T}) = \{(t, x)|0 \leq t \leq \bar{T}, 0 \leq x \leq L_i\}$ for equation (7.1) with the initial condition (7.5), the boundary condition (7.29) at $x = 0$ and the boundary condition (7.6) at $x = L_i$. By Lemma 2.1, this forward mixed problem admits a unique C^2 solution $u_i = u_i(t, x)$ with small C^2 norm on the domain $R_i(\bar{T})$. In particular, we have

$$\left| u_i(t, x) \right| + \left| \frac{\partial u_i}{\partial x}(t, x) \right| \leq \varepsilon_0, \quad \forall (t, x) \in R_i(\bar{T}) \quad (i = 2, \ldots, N - 1). \tag{7.30}$$

Thus, we can uniquely determine the value of u_{ix} $(i = 2, \ldots, N - 1)$ at $x = 0$ as

$$x = 0: \quad u_{ix} = b_i(t), \quad 0 \leq t \leq \bar{T} \quad (i = 2, \ldots, N - 1) \tag{7.31}$$

with small $C^1[0, \bar{T}]$ norm.

Noting (7.27) and (7.29), for each $i = 2, \ldots, N - 1$, by Lemma 2.1, it is easy to see that on the domain $R_i(T_2) = \{(t, x)|0 \leq t \leq T_2, 0 \leq x \leq L_i\}$, the C^2 solution $u_i = u_i(t, x)$ is just the solution $u_i = u_{if}(t, x)$ obtained in step (1), in particular, we have

$$b_i(t) \equiv \bar{b}_i(t), \quad 0 \leq t \leq T_2 \quad (i = 2, \ldots, N - 1). \tag{7.32}$$

(4) For $i = N$, noting (7.2), (7.25), (7.28) and (7.31), by using the interface condition (7.7), we can uniquely determine a function $b_N(t)$ such that the value of (u_N, u_{Nx}) at $x = 0$ can be given by

$$x = 0: \quad (u_N, u_{Nx}) = (a_1(t), b_N(t)), \quad 0 \le t \le \bar{T}. \tag{7.33}$$

Noting (7.27) and (7.32), it follows from the interface condition (7.7) that

$$b_N(t) \equiv \bar{b}_N(t), \quad 0 \le t \le T_2. \tag{7.34}$$

Considering the Nth string and exchanging the role of t and x, on the domain $R_N(\bar{T}) = \{(t, x) | 0 \le t \le \bar{T}, 0 \le x \le L_N\}$, we consider a rightward mixed initial-boundary value problem for equation (7.1) with the initial condition (7.33), the following boundary condition reduced from the original initial condition (7.5)

$$t = 0: \quad u_N = \varphi_N(x), \quad 0 \le x \le L_N \tag{7.35}$$

and the following artificial boundary condition

$$t = \bar{T}: \quad u_N = \tilde{\varphi}_N(x), \quad 0 \le x \le L_N, \tag{7.36}$$

where $\tilde{\varphi}_N(x)$ is a C^2 function with small $C^2[0, L_N]$ norm and satisfies the conditions of C^2 compatibility at the point $(t, x) = (\bar{T}, L_N)$.
By Lemma 2.1, this rightward mixed problem admits a unique C^2 solution $u_N = u_N(t, x)$ on the domain $R_N(\bar{T})$ with small C^2 norm. In particular, we have

$$\left| u_N(t, x) \right| + \left| \frac{\partial u_N}{\partial x}(t, x) \right| \le \varepsilon_0, \quad \forall (t, x) \in R_N(\bar{T}). \tag{7.37}$$

Both the C^2 solutions $u_N = u_N(t, x)$ and $u_N = u_{Nf}(t, x)$ satisfy simultaneously the same equation (7.1) (in which we take $i = N$), the same initial condition

$$x = 0: \quad (u_N, u_{Nx}) = (\bar{a}_1(t), \bar{b}_N(t)), \quad 0 \le t \le T_2 \tag{7.38}$$

and the same boundary condition (7.35). Noting (7.15), (7.17) and (7.37), by Lemma 2.3, the interval $\{0 \le x \le L_N\}$ on the initial axis $t = 0$ is included in the maximum determinate domain of the solution to this one-sided mixed problem, which means $u_N(t, x) \equiv u_{Nf}(t, x)$ on the interval $\{t = 0, 0 \le x \le L_N\}$, then $u_N = u_N(t, x)$ satisfies the initial condition (7.5).
Now, $(u_1(t, x), \ldots, u_N(t, x))$ is a desired piecewise C^2 solution. By substituting $u_N = u_N(t, x)$ into the boundary condition (7.6) at $x = L_N$, we get the desired boundary control function $h_N(t)$ for $0 \le t \le \bar{T}$.

Remark 7.1 In Theorems 7.1 and 7.2, the number of the given nodal profiles and that of the required boundary controls still satisfy the relationship (4.60). Here, 2 nodal profiles are given at the simple node E_1 and they should satisfy 1 constraint given by the boundary condition at this node, while, 1 boundary control is required.

7.3 Exact Boundary Controllability of Nodal Profile for 1-D Quasilinear Wave Equations on a Star-Like Network of Strings (Continued)

We now consider the case that the nodal profiles are given at the multiple node O on some strings, we have

Theorem 7.3 *Suppose that* $1 \le m < N$. *Let*

$$T > \max_{1 \le j \le m} \frac{L_j}{\sqrt{K_{jv}(0,0)}} \tag{7.39}$$

and $\bar{T}$ *be an arbitrarily given number such that*

$$\bar{T} > T. \tag{7.40}$$

For any given initial data $(\varphi_i(x), \psi_i(x))$ $(i = 1, \ldots, N)$ *and for any given boundary functions* $h_k(t)$ $(k = m+1, \ldots, N)$ *with small norms* $\|(\varphi_i(\cdot), \psi_i(\cdot))\|_{C^2[0,L] \times C^1[0,L]}$ $(i = 1, \ldots, N)$ *and* $\|h_k(\cdot)\|_{C^\delta[0,\bar{T}]}$ $(k = m+1, \ldots, N)$, *such that the conditions of* C^2 *compatibility are satisfied at the points* $(t, x) = (0, L_k)$ $(k = m+1, \ldots, N)$, *respectively, for any given values* $(\bar{\bar{u}}_j(t), \bar{\bar{v}}_j(t))$, *being the value of* (u_j, u_{jx}) *at* $x = 0$, *on the interval* $[T, \bar{T}]$ *with small norm* $\|(\bar{\bar{u}}_j(\cdot), \bar{\bar{v}}_j(\cdot))\|_{C^2[T,\bar{T}] \times C^1[T,\bar{T}]}$ $(j = 1, \ldots, m)$, *satisfying the corresponding interface condition* (7.8) *at the multiple node* O, *there exist boundary controls* $h_0(t)$ *and* $h_j(t)$ $(j = 1, \ldots, m)$ *with small norms* $\|h_0(\cdot)\|_{C^1[0,\bar{T}]}$ *and* $\|h_j(\cdot)\|_{C^\delta[0,\bar{T}]}$ $(j = 1, \ldots, m)$, *such that the mixed initial-boundary value problem* (7.1) *and* (7.5)–(7.8) *admits a unique piecewise* C^2 *solution* $u_i = u_i(t, x)$ $(i = 1, \ldots, N)$ *with small piecewise* C^2 *norm on the domain* $R(\bar{T}) = \bigcup_{i=1}^{N} R_i(\bar{T})$, *which fits exactly the given values* $(u_j, u_{jx}) = (\bar{\bar{u}}_j(t), \bar{\bar{v}}_j(t))$ $(j = 1, \ldots, m)$ *at the multiple node* O *for* $T \le t \le \bar{T}$ *(see Fig. 7.3).*

The proof of Theorem 7.3 is similar to the proof of Theorem 5.2 in Chap. 5.

Differently from **case 3** given by Theorem 7.3, and similarly to Theorem 7.2, in the case that all the nodal profiles are given at the multiple node O, we can only choose boundary functions at some simple nodes as controls (**case 4**), see Fig. 7.4.

Fig. 7.3 Exact boundary
controllability of nodal
profile on a star-like network
with nodal profiles given at
the multiple node (case 3)

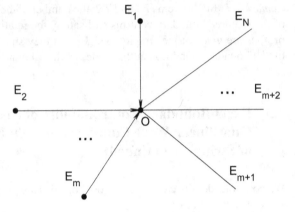

Fig. 7.4 Exact boundary
controllability of nodal
profile on a star-like network
with nodal profiles given at
the multiple node (case 4)

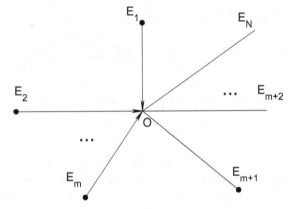

Theorem 7.4 *Suppose that* $1 \leq m < N$. *Let*

$$T > \max_{1 \leq l \leq m+1} \frac{L_l}{\sqrt{K_{lv}(0,0)}} \tag{7.41}$$

and $\bar{T}$ *be an arbitrarily given number such that*

$$\bar{T} > T. \tag{7.42}$$

For any given initial data $(\varphi_i(x), \psi_i(x))$ $(i = 1, \ldots, N)$ *and for any given functions* $h_0(t)$ *and* $h_k(t)$ $(k = m+2, \ldots, N)$ *with small norms* $\|(\varphi_i(\cdot), \psi_i(\cdot))\|_{C^2[0,L] \times C^1[0,L]}$ $(i = 1, \ldots, N)$, $\|h_0(\cdot)\|_{C^1[0,\bar{T}]}$ *and* $\|h_k(\cdot)\|_{C^\delta[0,\bar{T}]}$ $(k = m+2, \ldots, N)$, *such that the conditions of* C^2 *or piecewise* C^2 *compatibility are satisfied at the points* $(t, x) = (0, 0)$ *and* $(0, L_k)$ $(k = m+2, \ldots, N)$, *respectively, for any given value* $(\bar{\bar{u}}_j(t), \bar{\bar{v}}_j(t))$, *being the value of* (u_j, u_{jx}) *at* $x = 0$, *on the interval* $[T, \bar{T}]$ *with small norm* $\|(\bar{\bar{u}}_j(\cdot), \bar{\bar{v}}_j(\cdot))\|_{C^2[T,\bar{T}] \times C^1[T,\bar{T}]}$ *for each* $j = 1, \ldots, m$, *satisfying the corresponding interface condition (7.8) at the multiple node* O, *there*

exist boundary controls $h_l(t)$ with small norms $\|h_l(\cdot)\|_{C^\delta[0,\bar{T}]}$ for $l = 1, \ldots, m + 1$,
such that the mixed initial-boundary value problem (7.1) and (7.5)–(7.8) admits a
unique piecewise C^2 solution $u_i = u_i(t, x)$ $(i = 1, \ldots, N)$ with small piecewise
C^2 norm on the domain $R(\bar{T}) = \bigcup\limits_{i=1}^{N} R_i(\bar{T})$, which fits exactly the given values
$(u_j, u_{jx}) = (\bar{\bar{u}}_j(t), \bar{\bar{v}}_j(t))$ $(j = 1, \ldots, m)$ at the multiple node O for $T \le t \le \bar{T}$.

Proof We still prove it by a constructive method.

(1) By (7.41), there exists an $\varepsilon_0 > 0$ so small that

$$T_1 \overset{\text{def.}}{=} \max_{1 \le l \le m+1} \max_{|u_l|+|v_l| \le \varepsilon_0} \frac{L_l}{\sqrt{K_{lv}(u_l, v_l)}} < T. \tag{7.43}$$

For this star-like network of strings, on the domain $R(T_1) = \bigcup\limits_{i=1}^{N} R_i(T_1) = \bigcup\limits_{i=1}^{N}\{(t, x)|0 \le t \le T_1, 0 \le x \le L_i\}$, we consider a forward mixed initial-boundary value problem for equation (7.1) with the initial condition (7.5), the interface conditions (7.7)–(7.8) at the multiple node O, the boundary conditions (7.6) at the simple node E_k $(k = m + 2, \ldots, N)$ and the following artificial boundary conditions at the simple node E_l $(l = 1, \ldots, m + 1)$

$$x = L_l : \quad u_l = f_l(t) \quad (l = 1, \ldots, m + 1), \tag{7.44}$$

where $f_l(t)$ $(l = 1, \ldots, m+1)$ are C^2 functions with small $C^2[0, T_1]$ norms, such that the conditions of C^2 compatibility are satisfied at the points $(t, x) = (0, L_l)$ $(l = 1, \ldots, m + 1)$, respectively.

By Lemma 3.2, this forward mixed problem admits a unique piecewise C^2 solution $u_i = u_{if}(t, x)$ $(i = 1, \ldots, N)$ with small piecewise C^2 norm on the domain $R(T_1) = \bigcup\limits_{i=1}^{N} R_i(T_1)$. In particular, we have

$$\left|u_{if}(t, x)\right| + \left|\frac{\partial u_{if}}{\partial x}(t, x)\right| \le \varepsilon_0, \quad \forall(t, x) \in R_i(T_1) \quad (i = 1, \ldots, N). \tag{7.45}$$

Thus, we can uniquely determine the corresponding values of $\left(u_{if}, \frac{\partial u_{if}}{\partial x}\right)$ $(i = 1, \ldots, N)$ at $x = 0$ as

$$x = 0 : \quad \left(u_{if}, \frac{\partial u_{if}}{\partial x}\right) = (\bar{u}_i(t), \bar{v}_i(t)), \quad 0 \le t \le T_1 \quad (i = 1, \ldots, N) \tag{7.46}$$

with small $C^2[0, T_1] \times C^1[0, T_1]$ norm.

Noting (7.43), there exist $C^2 \times C^1$ vector functions $(u_j(t), v_j(t))$ $(j = 1, \ldots, m)$ with small $C^2[0, \bar{T}] \times C^1[0, \bar{T}]$ norm, such that

$$(u_j(t), v_j(t)) = \begin{cases} (\bar{u}_j(t), \bar{v}_j(t)), & 0 \le t \le T_1, \\ (\bar{\bar{u}}_j(t), \bar{\bar{v}}_j(t)), & T \le t \le \bar{T}, \end{cases} \quad (j = 1, \ldots, m) \quad (7.47)$$

and $(u_j(t), v_j(t))$ $(j = 1, \ldots, m)$ satisfy the corresponding interface condition (7.8) on the whole interval $0 \le t \le \bar{T}$, namely, we have

$$u_1(t) \equiv \ldots \equiv u_m(t), \quad 0 \le t \le \bar{T}. \quad (7.48)$$

(2) By the interface condition (7.8) at $x = 0$, noting (7.48), we have

$$x = 0: \quad u_{m+1} = \cdots = u_N = u_1(t), \quad 0 \le t \le \bar{T}, \quad (7.49)$$

in particular, we have

$$x = 0: \quad u_{m+2} = \cdots = u_N = u_1(t), \quad 0 \le t \le \bar{T}. \quad (7.50)$$

For each $k = m+2, \ldots, N$, we consider a forward mixed initial-boundary value problem on the domain $R_k(\bar{T}) = \{(t, x) | 0 \le t \le \bar{T}, 0 \le x \le L_k\}$ for equation (7.1) with the initial condition (7.5), the boundary condition (7.50) at $x = 0$ and the boundary condition (7.6) at $x = L_k$. By Lemma 2.1, this forward mixed problem admits a unique C^2 solution $u_k = u_k(t, x)$ with small $C^2(R_k(\bar{T}))$ norm for $k = m + 2, \ldots, N$. In particular, we have

$$\left| u_k(t, x) \right| + \left| \frac{\partial u_k}{\partial x}(t, x) \right| \le \varepsilon_0, \quad \forall (t, x) \in R_k(\bar{T}) \quad (k = m + 2, \ldots, N). \quad (7.51)$$

Thus, we can uniquely determine the values of u_{kx} $(k = m+2, \ldots, N)$ at $x = 0$ as

$$x = 0: \quad u_{kx} = v_k(t), \quad 0 \le t \le \bar{T} \quad (k = m + 2, \ldots, N) \quad (7.52)$$

with small $C^1[0, \bar{T}]$ norm. In addition, it is easy to see that

$$v_k(t) \equiv \bar{v}_k(t), \quad 0 \le t \le T_1 \quad (k = m + 2, \ldots, N). \quad (7.53)$$

(3) Noting (7.2), by (7.47), (7.49) and the interface conditions (7.7)–(7.8), we can uniquely determine a function $v_{m+1}(t)$ such that the value of $(u_{m+1}, u_{m+1,x})$ at $x = 0$ is given by

$$x = 0: \quad (u_{m+1}, u_{m+1,x}) = (u_1(t), v_{m+1}(t)), \quad 0 \le t \le \bar{T}, \qquad (7.54)$$

moreover, $v_{m+1}(t) \equiv \bar{v}_{m+1}(t)$ for $0 \le t \le T_1$.

For each $l = 1, \ldots, m + 1$, considering the lth string and exchanging the role of t and x, on the domain $R_l(\bar{T}) = \{(t, x) | 0 \le t \le \bar{T}, 0 \le x \le L_l\}$, we solve a rightward mixed initial-boundary value problem for equation (7.1) with the initial condition

$$x = 0: \quad (u_l, u_{lx}) = (u_1(t), v_l(t)), \quad 0 \le t \le \bar{T} \quad (l = 1, \ldots, m + 1),$$
$$(7.55)$$

the boundary condition at $t = 0$ reduced from the original initial condition (7.5)

$$t = 0: \quad u_l = \varphi_l(x), \quad 0 \le x \le L_l \quad (l = 1, \ldots, m + 1) \qquad (7.56)$$

and the following artificial boundary condition

$$t = \bar{T}: \quad u_l = \tilde{\varphi}_l(x), \quad 0 \le x \le L_l \quad (l = 1, \ldots, m + 1), \qquad (7.57)$$

where $\tilde{\varphi}_l(x)$ $(l = 1, \ldots, m + 1)$ are C^2 functions with small $C^2[0, L_l]$ norms, such that the conditions of C^2 compatibility are satisfied at the points $(t, x) = (\bar{T}, L_l)$ $(l = 1, \ldots, m + 1)$, respectively.

For each $l = 1, \ldots, m + 1$, this rightward mixed problem admits a unique C^2 solution $u_l = u_l(t, x)$ on the domain $R_l(\bar{T})$ with small C^2 norm. In particular, we have

$$\left| u_l(t, x) \right| + \left| \frac{\partial u_l}{\partial x}(t, x) \right| \le \varepsilon_0, \quad \forall (t, x) \in R_l(\bar{T}) \quad (l = 1, \ldots, m + 1). \quad (7.58)$$

Both the C^2 solutions $u_l = u_l(t, x)$ and $u_l = u_{lf}(t, x)$ satisfy simultaneously the same equation (7.1), the same initial condition

$$x = 0: \quad (u_l, u_{lx}) = (u_1(t), v_l(t)), \quad 0 \le t \le T_1 \qquad (7.59)$$

and the same boundary condition (7.56) for $l = 1, \ldots, m + 1$. Noting (7.43), (7.45) and (7.58), by Lemma 2.3, the interval $\{0 \le x \le L_l\}$ on the initial axis $t = 0$ is included in the maximum determinate domain of the solution to this one-sided mixed problem, which means $u_l(t, x) \equiv u_{lf}(t, x)$ on the interval $\{t = 0, 0 \le x \le L_l\}$, then $u_l = u_l(t, x)$ satisfies the initial condition (7.5) for $l = 1, \ldots, m + 1$.

Now, $(u_1(t, x), \ldots, u_N(t, x))$ is a required piecewise C^2 solution to this problem. For each $l = 1, \ldots, m + 1$, by substituting $u_l = u_l(t, x)$ into the boundary condition (7.6) at $x = L_l$, we get the desired boundary controls $h_l(t)$ $(l = 1, \ldots, m + 1)$ for $0 \le t \le \bar{T}$.

Remark 7.2 In Theorems 7.3 and 7.4, the number of the given nodal profiles and the number of the required boundary controls still satisfy the relationship (4.60). Here, $2m$ nodal profiles are provided at the multiple node O, which satisfy $(m - 1)$ constraints given by the interface condition (7.8), while, the number of the desired boundary controls is $m + 1$.

Corresponding to the missing case $m = N$ in Theorems 7.3 and 7.4, we now consider the case that the nodal profiles are given at the multiple node O on all the strings, while, the boundary controls are chosen at all the simple nodes $x = L_i$ $(i = 1, \ldots, N)$ (**case 5**). We have

Theorem 7.5 *Let*

$$T > \max_{1 \le i \le N} \frac{L_i}{\sqrt{K_{iv}(0, 0)}} \tag{7.60}$$

and $\bar{T}$ be an arbitrarily given number such that

$$\bar{T} > T. \tag{7.61}$$

For any given initial data $(\varphi_i(x), \psi_i(x))$ $(i = 1, \ldots, N)$ and for any given total stress function $h_0(t)$ with small norms $\|(\varphi_i(\cdot), \psi_i(\cdot))\|_{C^2[0,L] \times C^1[0,L]}$ $(i = 1, \ldots, N)$ and $\|h_0(\cdot)\|_{C^1[0,\bar{T}]}$, such that the conditions of piecewise C^2 compatibility are satisfied at the point $(t, x) = (0, 0)$, for any given $(\bar{\bar{u}}_i(t), \bar{\bar{v}}_i(t))$, being the value of (u_i, u_{ix}), on the interval $[T, \bar{T}]$ on the ith string, with small norm $\|(\bar{\bar{u}}_i(\cdot), \bar{\bar{v}}_i(\cdot))\|_{C^2[T,\bar{T}] \times C^1[T,\bar{T}]}$ for $i = 1, \ldots, N$, satisfying the interface conditions (7.7)–(7.8) at the multiple node O, there exist boundary controls $h_i(t)$ $(i = 1, \ldots, N)$ with small norm $\|h_i(\cdot)\|_{C^5[0,\bar{T}]}$ $(i = 1, \ldots, N)$, such that the mixed initial-boundary value problem (7.1) and (7.5)–(7.8) admits a unique piecewise C^2 solution $u_i = u_i(t, x)$ $(i = 1, \ldots, N)$ with small piecewise C^2 norm on the domain $R(\bar{T}) = \bigcup_{i=1}^{N} R_i(\bar{T})$, which fits exactly the given values $(u_i, u_{ix}) = (\bar{\bar{u}}_i(t), \bar{\bar{v}}_i(t))$ $(i = 1, \ldots, N)$ at the multiple node O for $T \le t \le \bar{T}$ (see Fig. 7.5).

The proof of Theorem 7.5 is similar to that of Theorem 5.3.

Remark 7.3 In Theorem 7.5, the number of the given nodal profiles and the number of the required boundary controls still satisfy the relationship (4.60). Here, $2N$ nodal profiles are provided at the multiple node O, which satisfy N constraints given by the interface conditions (7.7)–(7.8), while, the number of the desired boundary controls is N.

Fig. 7.5 Exact boundary
controllability of nodal
profile on a star-like network
with nodal profiles given at
the multiple node (case 5)

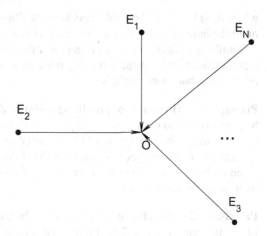

7.4 Exact Boundary Controllability of Nodal Profile for Quasilinear Wave Equations on a Planar Tree-Like Network of Strings

In this section, we generalize the results given in Sects. 7.2 and 7.3 to the case of a planar tree-like network composed of N strings: $C_1, \ldots, C_N$.

Without loss of generality, we suppose that one endpoint of string C_1 is a simple node in this network, and we take it as a starting node E. For the ith string, let d_{i0} and d_{i1} be the x coordinates of its two endpoints, and its length be $L_i = d_{i1} - d_{i0}$. For simplicity, in what follows, we simply use d_{i0} (resp. d_{i1}) to express the node $x = d_{i0}$ (resp. $x = d_{i1}$). We suppose that d_{i0} is closer to E than d_{i1} in this network and d_{10} is just E.

Let $\mathcal{M}$ and $\mathcal{S}$ be two subsets of $\{1, \ldots, N\}$, such that $i \in \mathcal{M}$ if and only if d_{i1} is a multiple node, while, $i \in \mathcal{S}$ if and only if d_{i1} is a simple node. In other words, $\mathcal{M}$ is the set of indices corresponding to all the multiple nodes, while $\mathcal{S}$ is the set of indices corresponding to all the simple nodes except E.

On this network, we still consider the quasilinear wave equations (7.1) with the initial condition (7.5), the boundary condition (7.6) at any simple node d_{10} or d_{i1} ($i \in \mathcal{S}$) and the following interface conditions at any multiple node d_{i1} ($i \in \mathcal{M}$):

$$\sum_{j \in \mathcal{J}_i} K_j(u_j, u_{jx}) + K_i(u_i, u_{ix}) = h_{i0}(t) \tag{7.62}$$

and

$$u_j = u_i, \quad \forall j \in \mathcal{J}_i, \tag{7.63}$$

where $h_{i0}(t) \in C^1$ is the total stress at the multiple node d_{i1}, and $\mathcal{J}_i$ denotes the set of all the indices j such that the node d_{j0} is just the node d_{i1}.

Similarly to Sect. 5.6, for a given planar tree-like network of strings with general topology, we first show the following principles of providing nodal profiles and that of choosing boundary controls.

Principle 7.1 (Principle of providing nodal profiles) Nodal profiles should always be given as a form of $(u_i, u_{ix}) = (\bar{\bar{u}}_i(t), \bar{\bar{v}}_i(t))$. For any given string in this tree-like network, the nodal profiles should be possibly given only at one node of this string. For any given two strings with a joint node, if the nodal profiles are simultaneously given on both strings, then these two pairs of nodal profiles should be given at the joint node (multiple node).

Principle 7.2 (Principle of choosing boundary controls) For any given nodal profiles satisfying Principle 7.1, if the nodal profiles are given at a simple node, then at another node (multiple node) of the corresponding string, we choose $h_{i0}(t)$ in the corresponding interface condition (7.62) as a boundary control. If some nodal profiles are given at a multiple node, then (1) when these nodal profiles are not given on all the strings with this multiple node as their common node, we choose $h_{i0}(t)$ in the interface condition (7.62) at this multiple node as a boundary control, meanwhile, we choose boundary function as a control at another node (simple or multiple node) of the string corresponding to the given nodal profiles. (2) when these nodal profiles are given on all the strings with this multiple node as their common node, we choose $h_i(t)$ in (7.6) at all other nodes of these strings as boundary controls.

Based on Principles 7.1 and 7.2, we can get the following theorem.

Theorem 7.6 *For any given planar tree-like network of strings, there exists a suitably large constant $T > 0$, such that for any given constant $\bar{T}$ satisfying $\bar{T} > T$, if some nodal profiles are given according to Principle 7.1 on the interval $[T, \bar{T}]$, then we can choose corresponding controls according to Principle 7.2 on the interval $[0, \bar{T}]$, such that the exact boundary controllability of nodal profile can be realized on this tree-like network, and the number of the given nodal profiles and the number of the required boundary controls satisfy the relationship (4.60) (see Fig. 7.6).*

Here, we always assume that the C^2 or C^1 norms of the initial functions, the boundary functions and the nodal profiles are suitably small and the corresponding conditions of C^2 or piecewise C^2 compatibility are satisfied.

Proof For any given tree-like network of strings, we prove Theorem 7.6 by an induction on the number of multiple nodes, at which there are some given nodal profiles satisfying Principle 7.1.

First, we assume that the number of the multiple nodes with given nodal profiles is 0, namely, the nodal profiles are only given at simple nodes. Without loss of generality, suppose that the nodal profiles $(\bar{\bar{u}}_i(t), \bar{\bar{v}}_i(t))$ are given at some simple nodes d_{i1} $(i \in \mathcal{S})$, respectively. According to Principle 7.1, these strings C_i corresponding to d_{i1} do not have a common node with each other. Cutting down these C_i from the

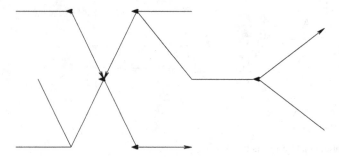

Fig. 7.6   Exact boundary controllability of nodal profile on a tree-like network of strings

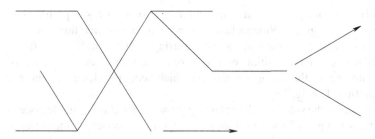

Fig. 7.7   Splitting a tree-like network

original tree-like network, such that the corresponding node d_{i0} becomes a simple node on the left subnetworks (see Fig. 7.7). All the left subnetworks do not include any nodal profiles. By Theorem 6.1 for the exact boundary controllability of nodal profile on a single string, we can construct a C^2 solution on each such string C_i, which satisfies the wave equation (7.1), the initial condition (7.5) and the given nodal profiles $(\bar{\bar{u}}_i(t), \bar{\bar{v}}_i(t))$ on a suitable time interval at d_{i1}. Then, by the interface condition (7.63), we can obtain the values of u at the corresponding joint nodes of the left subnetworks with C_i, respectively. Taking these values as the boundary conditions at the corresponding simple nodes of the left subnetworks, then the wave equations (7.1) with the initial condition (7.5) and the corresponding boundary conditions and interface conditions construct a well-posed forward mixed initial-boundary value problem on each left subnetwork, which admits a piecewise C^2 solution. Substituting all the C^2 solutions on these C_i and all the piecewise C^2 solutions on the left subnetworks into the interface condition (7.62), we obtain the total stress functions at corresponding joint nodes, which can be taken as the boundary controls and satisfy Principle 7.2.

Suppose that the exact boundary controllability of nodal profile holds for a tree-like network with n multiple nodes, on which the nodal profiles satisfying Principle 7.1 are given. We now consider a tree-like network including $(n + 1)$ multiple nodes, on which the nodal profiles satisfying Principle 7.1 are given, we want to prove that the exact boundary controllability of nodal profile still holds in this case. Let d_{i1}

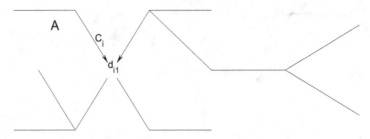

Fig. 7.8 Splitting a tree-like network

$(i \in \mathcal{M})$ be a multiple node, at which p ($p \geq 1$) pairs of nodal profiles satisfying Principle 7.1 are given. Without loss of generality, we assume that a pair of nodal profiles $(\bar{\bar{u}}_i(t), \bar{\bar{v}}_i(t))$ at the node d_{i1} of the string C_i is given. Splitting the original tree-like network at the multiple node d_{i1}, we get several subnetworks. It is obvious that the number of the multiple nodes, on which nodal profiles are given, is strictly less than $(n + 1)$ (Fig. 7.8).

Let A denote the subnetwork containing the string C_i. d_{i1} is a simple node on A, at which there is a pair of nodal profiles $(\bar{\bar{u}}_i(t), \bar{\bar{v}}_i(t))$. By induction, the corresponding exact boundary controllability of nodal profile on the subnetwork A holds, that means we can construct a piecewise C^2 solution on A, which fits exactly the given nodal profile, then, the value of (u_i, u_{ix}) at $x = d_{i1}$ on the whole time interval is determined.

For the subnetworks that do not contain the string C_i, by the interface condition (7.63), we can determine the values of u at the node d_{i1} on these subnetworks, moreover, by Principle 7.1, if the nodal profiles are given on some strings with d_{i1} as an end of them, then these nodal profiles should only be given at d_{i1}. By induction, the corresponding exact boundary controllability of nodal profile on these subnetworks holds, then the mixed problem of the wave equations (7.1) with the initial condition (7.5), the corresponding boundary conditions and interface conditions admits a piecewise C^2 solution on any one of these subnetworks, which fits exactly the given nodal profiles, respectively. Put all these piecewise C^2 solutions together with the piecewise C^2 solution on A, we get a desired piecewise C^2 solution for the original tree-like network. Substituting it into the interface condition (7.62) at the multiple node d_{i1}, we obtain a corresponding total stress function, which is the desired boundary control function at this node.

Finally, according to Principles 7.1 and 7.2, it is easy to see that the number of the given nodal profiles and that of the desired controls still satisfy the relationship (4.60). This finishes the proof.

Principle 7.3 (Principle of choosing and transferring boundary controls) The boundary controls chosen according to Principle 7.2 may be transferred at multiple nodes. For example, if we have chosen $h_{i0}(t)$ as the boundary control at the multiple node d_{i1}, and if there exists a connected string-like network starting from d_{i1}, which does not contain any nodal profile or boundary control, then at any node in this string-

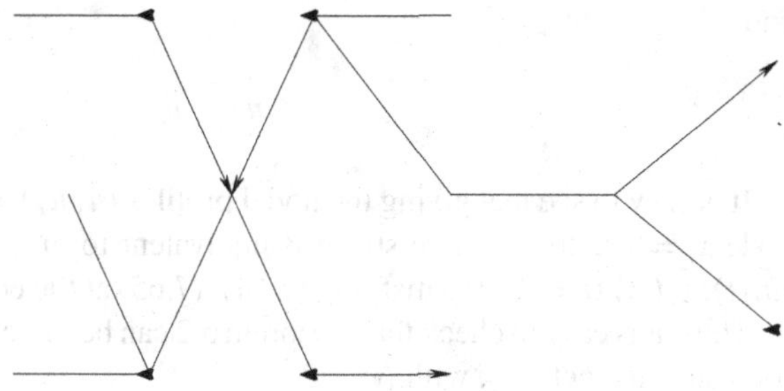

Fig. 7.9   Transfer of
boundary controls (1)

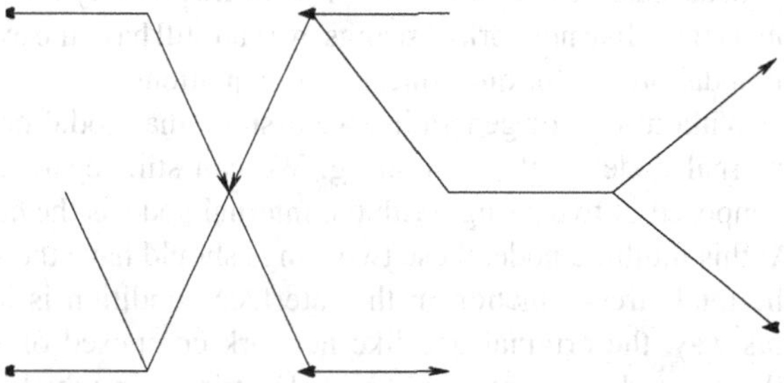

Fig. 7.10   Transfer of
boundary controls (2)

like network, the corresponding boundary (or total stress) function can be used to replace $h_{i0}(t)$ as the new boundary control function.

Different controls at different multiple nodes can be transferred simultaneously according to Principle 7.3, however, the corresponding string-like networks should not intersect with each other in the transferring process, see Figs. 7.9 and 7.10.

7.5   Exact Boundary Controllability of Nodal Profile for Quasilinear Wave Equations on a Planar Tree-Like Network (Continued)

In this section, corresponding to the result given in Theorem 6.2, we discuss the case that nodal profiles are given at some internal nodes.

Similarly to Sect. 5.7, we first reconsider Theorem 6.2 from a different point of view,

The single string $[0, L]$ with an internal node $x = \theta L$ $(0 < \theta < 1)$ can be artificially regarded as a star-like network composed of two strings by taking this internal node as the multiple node, and at this multiple node these two strings should have the same function $K_i(u, v) \overset{\text{def.}}{=} K(u, v)$ $(i = 1, 2)$, moreover, the total stress function $h_0(t)$ should be identically equal to zero. Thus, the interface conditions (7.7)–(7.8) at this multiple node can be written as

$$\sum_{i=1}^{2} K(u_i, u_{ix}) = 0 \tag{7.64}$$

and

$$u_2 = u_1. \tag{7.65}$$

It is easy to see that giving the nodal profiles $(u, u_x) = (\bar{\bar{u}}(t), \bar{\bar{v}}(t))$ at the internal node $x = \theta L$ for a single string is equivalent to give two pairs of nodal profiles $(\bar{\bar{u}}_i(t), \bar{\bar{v}}_i(t))$ $(i = 1, 2)$ satisfying (7.64)–(7.65) at the corresponding multiple node.

Thus, it is easy to check that Theorem 6.2 can be directly obtained by Theorem 7.5 for a star-like network with $N = 2$.

In the case that certain nodal profiles are possibly given at some internal nodes on a planar tree-like network of strings, we can still have the exact boundary controllability of nodal profile for quasilinear wave equations.

Without loss of generality, we assume that nodal profiles are given only at one internal node on the first string. We can still regard this string as a subnetwork composed of two strings with the internal node as the corresponding multiple node. At this multiple node, these two stings should have the same function $G(u, u_x)$ and the total stress function in the interface condition is identically equal to zero. In this way, the original tree-like network composed of N strings turns into a tree-like network composed of $(N + 1)$ strings, in which two pairs of nodal profiles $(\bar{\bar{u}}_i(t), \bar{\bar{v}}_i(t))$ $(i = 1, 2)$ satisfying the interface conditions (7.64)–(7.65) are given from both sides of the corresponding multiple node.

When these two pairs of nodal profiles satisfy Principle 7.1, by means of the corresponding boundary controls satisfying Principles 7.2 and 7.3, we can still get the exact boundary controllability of nodal profile for the original tree-like network with nodal profiles given at an internal node.

However, when these two pairs of nodal profiles do not satisfy Principle 7.1, for instance, there are already nodal profiles given at one of the two nodes of the corresponding string, it is impossible to provide nodal profiles at this internal node, and we can not get the exact boundary controllability of nodal profile on the original tree-like network with nodal profiles given at this internal node.

Remark 7.4 In the case discussed in this section, relationship (4.60) still holds.

7.6 Remarks

For general 1-D quasilinear hyperbolic equations (systems) of second order, the exact boundary controllability of nodal profile can be discussed in a similar way in principle, provided that there are no zero eigenvalues. However, the physically meaningful interface conditions at the multiple nodes for each such model should be carefully presented and considered, and then a precise study is worth to be done in each case.

When there are zero eigenvalues, the whole problem remains still open.

References

1. Bressan, A., Čanić, S., Garavello, M., Herty, M., Piccoli, B.: Flows on networks: recent results and perspectives. EMS Surv. Math. Sci. **1**, 47–111 (2014)
2. Coron, J.-M., Wang, Z.: Controllability for a scalar conservation law with nonlocal velocity. J. Differ. Equ. **252**, 181–201 (2012)
3. D'Apice, C., Göttlich, S., Herty, M., Piccoli, B.: Modeling, Simulation, and Optimization of Supply Chains: A Continuous Approach, Society for Industrial and Applied Mathematics (SIAM). Philadelphia, PA (2010)
4. de Saint-Venant, B.: Théorie du mouvement non permanent des eaux, avec application aux crues des rivières et l'introduction des marées dans leur lit, C. R. Acad. Sci. **73**, 147–154, 237–240 (1871)
5. Garavello, M., Piccoli, B.: Traffic flow on networks. AIMS Ser. Appl. Math. **1** (2006). American Institute of Mathematical Sciences (AIMS), Springfield, MO
6. GU, Q., Li, T.: Exact boundary controllability for quasilinear wave equations in a planar tree-like network of strings. Ann. de L'Institut Henri Poincaré Anal. Non Linéaire **26**, 2373–2384 (2009)
7. GU, Q., Li, T.: Exact boundary controllability for quasilinear hyperbolic systems on a tree-like network and its applications. SIAM J. Control Optim. **49**, 1404–1421 (2011)
8. Gu, Q., Li, T.: Exact boundary controllability of nodal profile for quasilinear hyperbolic systems in a tree-like network. Math. Methods Appl. Sci. **34**, 911–928 (2011)
9. Gu, Q., Li, T.: Exact boundary controllability of nodal profile for unsteady flows on a tree-like network of open canals. J. de Mathématiques Pures et Appliquées **99**, 86–105 (2013)
10. Gugat, M., Herty, M., Schleper, V.: Flow control in gas networks: exact controllability to a given demand. Math. Methods Appl. Sci. **34**, 745–757 (2011)
11. Li, T.: Controllability and observability: From ODEs to quasilinear hyperbolic systems. In: Jeltsch, R., Wanner, G. (eds.) Sixth International Congress on Industrial and Applied Mathematics (ICIAM 07), Zürich, Switzerland, 16–20 July 2007. Invited Lectures, European Mathematical Society, 2009, 251–278
12. Li, T.: Controllability and observability for quasilinear hyperbolic systems. AIMS Series on Applied Mathematics, vol. 3. American Institute of Mathematical Sciences & Higher Education Press (2010)
13. Li, T.: Exact boundary controllability of nodal profile for quasilinear hyperbolic systems. Math. Methods Appl. Sci. **33**, 2101–2106 (2010)
14. Li, T.: A constructive method to controllability and observability for quasilinear hyperbolic systems. Methods Appl. Anal. **18**, 69–84 (2011)
15. Li, T., Jin, Y.: Semi-global C^1 solution to the mixed initial-boundary value problem for quasilinear hyperbolic systems. Chin. Ann. Math. Ser. B **22**, 325–336 (2001)

© The Author(s) 2016

T. Li et al., *Exact Boundary Controllability of Nodal Profile*
for Quasilinear Hyperbolic Systems, SpringerBriefs in Mathematics,
DOI 10.1007/978-981-10-2842-7

16. Li, T., Rao, B.: Local exact boundary controllability for a class of quasilinear hyperbolic systems. Chin. Ann. Math. Ser. B **23**, 209–218 (2002)
17. Li, T., Rao, B.: Exact boundary controllability for quasilinear hyperbolic systems. SIAM J. Control Optim. **41**, 1748–1755 (2003)
18. Li, T., Rao, B.: Strong (weak) exact controllability and strong (weak) exact observability for quasilinear hyperbolic systems. Chin. Ann. Math. Ser. B **31**, 723–742 (2010)
19. Li, T., Yu, L.: Exact boundary controllability for 1-D quasilinear wave equations. SIAM J. Control Optim. **45**, 1074–1083 (2006)
20. Li, T., Yu, W.: Boundary Value Problems for Quasilinear Hyperbolic Systems. Duke University Mathematics Series V (1985)
21. Wang, K.: Exact boundary controllability of nodal profile for 1-D quasilinear wave equations. Frontiers Math. China **6**, 545–555 (2011)
22. Wang, K., Gu, Q.: Exact boundary controllability of nodal profile for quasilinear wave equations in a planar tree-like network of strings. Math. Methods Appl. Sci. **37**, 1206–1218 (2014)
23. Zhuang, K.: Exact controllability of nodal profile for 1-D first order quasi-linear hyperbolic systems with zero eigenvalues, Preprint
24. Zhuang, K.: Exact controllability of nodal profile with internal controls for 1-D first order quasilinear hyperbolic systems, Preprint

Index

© The Author(s) 2016
T. Li et al., *Exact Boundary Controllability of Nodal Profile for Quasilinear Hyperbolic Systems*, SpringerBriefs in Mathematics, DOI 10.1007/978-981-10-2842-7

Printed in the United States
By Bookmasters